1990
YEARBOOK of
ASTRONOMY

1990
YEARBOOK of
ASTRONOMY

edited by

Patrick Moore

W. W. Norton & Company

NEW YORK   LONDON

ISBN 0-393-02781-3

W. W. Norton & Company, Inc.,
500 Fifth Avenue, New York, NY 10110
W. W. Norton & Company, Limited,
37 Great Russell Street, London WC1B 3NU

Printed in Great Britain

1 2 3 4 5 6 7 8 9 0

Contents

Editor's Foreword

Our 28th *Yearbook* follows the usual, well-established pattern; since it seems to be well received, I see no reason to change it. In addition to providing the lists of monthly phenomena, Gordon Taylor has also contributed a major article this year – and the subject of timekeeping is of prime importance to us all.

We welcome, once again, our valued regular contributors, Dr David Allen and Dr Paul Murdin; we are delighted to have an article from Professor Michael Disney and his team from Cardiff, particularly since it deals with a new branch of observational astronomy which may well prove to be of extreme importance. In view of this year's total solar eclipse, the advice given by Michael Maunder, one of our leading eclipse photographers, is timely.

My own contribution, in addition to the monthly notes, is concerned with the Voyager pass of Neptune. As this occurred in August 1989, and we have to go to press in a matter of days after that, I have had to be brief; but there will be a more detailed account in the 1991 *Yearbook*.

PATRICK MOORE

Selsey, April 1989

Preface

New readers will find that all the information in this *Yearbook* is given in diagrammatic or descriptive form; the positions of the planets may easily be found on the specially designed star charts, while the monthly notes describe the movements of the planets and give details of other astronomical phenomena visible in both the northern and southern hemispheres. Two sets of the star charts are provided. The **Northern Charts** (pp. 14 to 39) are designed for use in latitude 52 degrees north, but may be used without alteration throughout the British Isles, and (except in the case of eclipses and occultations) in other countries of similar north latitude. The **Southern Charts** (pp. 40 to 65) are drawn for latitude 35 degrees south, and are suitable for use in South Africa, Australia and New Zealand, and other stations in approximately the same south latitude. The reader who needs more detailed information will find *Norton's Star Atlas* (Longman) an invaluable guide, while more precise positions of the planets and their satellites, together with predictions of occultations, meteor showers, and periodic comets may be found in the *Handbook* of the British Astronomical Association. The British monthly periodical, with current news, articles, and monthly notes is *Astronomy Now*. Readers will also find details of forthcoming events given in the American *Sky and Telescope*. This monthly publication also produces a special occultation supplement giving predictions for the United States and Canada.

Important Note
The times given on the star charts and in the Monthly Notes are generally given as local times, using the 24-hour clock, the day beginning at midnight. All the dates, and the times of a few events (e.g. eclipses), are given in Greenwich Mean Time (G.M.T.), which is related to local time by the formula

Local Mean Time = G.M.T. − west longitude

In practice, small differences of longitudes are ignored, and the observer will use local clock time, which will be the appropriate

9

Standard (or Zone) Time. As the formula indicates, places in west longitude will have a Standard Time slow on G.M.T., while places in east longitude will have a Standard Time fast on G.M.T. As examples we have:

Standard Time in

New Zealand	G.M.T.	+	12 hours
Victoria; N.S.W.	G.M.T.	+	10 hours
Western Australia	G.M.T.	+	8 hours
South Africa	G.M.T.	+	2 hours
British Isles	G.M.T.		
Eastern S.T.	G.M.T.	−	5 hours
Central S.T.	G.M.T.	−	6 hours, etc.

If Summer Time is in use, the clocks will have to have been advanced by one hour, and this hour must be subtracted from the clock time to give Standard Time.

In Great Britain and N. Ireland, Summer Time will be in force in 1990 from March 25^{d}01^{h} until October 28^{d}01^{h} G.M.T.

Notes on the Star Charts

The stars, together with the Sun, Moon and planets seem to be set on the surface of the celestial sphere, which appears to rotate about the Earth from east to west. Since it is impossible to represent a curved surface accurately on a plane, any kind of star map is bound to contain some form of distortion. But it is well known that the eye can endure some kinds of distortion better than others, and it is particularly true that the eye is most sensitive to deviations from the vertical and horizontal. For this reason the star charts given in this volume have been designed to give a true representation of vertical and horizontal lines, whatever may be the resulting distortion in the shape of a constellation figure. It will be found that the amount of distortion is, in general, quite small, and is only obvious in the case of large constellations such as Leo and Pegasus, when these appear at the top of the charts, and so are drawn out sideways.

The charts show all stars down to the fourth magnitude, together with a number of fainter stars which are necessary to define the shape of a constellation. There is no standard system for representing the outlines of the constellations, and triangles and other simple figures have been used to give outlines which are easy to follow with the naked eye. The names of the constellations are given, together with the proper names of the brighter stars. The apparent magnitudes of the stars are indicated roughly by using four different sizes of dots, the larger dots representing the brighter stars.

The two sets of star charts are similar in design. At each opening there is a group of four charts which give a complete coverage of the sky up to an altitude of 62½ degrees; there are twelve such groups to cover the entire year. In the **Northern Charts** (for 52 degrees north) the upper two charts show the southern sky, south being at the centre and east on the left. The coverage is from 10 degrees north of east (top left) to 10 degrees north of west (top right). The two lower charts show the northern sky from 10 degrees south of west (lower left) to 10 degrees south of east (lower right). There is thus an overlap east and west.

Conversely, in the **Southern Charts** (for 35 degrees south) the upper two charts show the northern sky, with north at the centre and east on the right. The two lower charts show the southern sky, with south at the centre and east on the left. The coverage and overlap is the same on both sets of charts.

Because the sidereal day is shorter than the solar day, the stars appear to rise and set about four minutes earlier each day, and this amounts to two hours in a month. Hence the twelve groups of charts in each set are sufficient to give the appearance of the sky throughout the day at intervals of two hours, or at the same time of night at monthly intervals throughout the year. The actual range of dates and times when the stars on the charts are visible is indicated at the top of each page. Each group is numbered in bold type, and the number to be used for any given month and time is summarized in the following table:

Local Time	18h	20h	22h	0h	2h	4h	6h
January	11	12	1	2	3	4	5
February	12	1	2	3	4	5	6
March	1	2	3	4	5	6	7
April	2	3	4	5	6	7	8
May	3	4	5	6	7	8	9
June	4	5	6	7	8	9	10
July	5	6	7	8	9	10	11
August	6	7	8	9	10	11	12
September	7	8	9	10	11	12	1
October	8	9	10	11	12	1	2
November	9	10	11	12	1	2	3
December	10	11	12	1	2	3	4

The charts are drawn to scale, the horizontal measurements, marked at every 10 degrees, giving the azimuths (or true bearings) measured from the north round through east (90 degrees), south (180 degrees), and west (270 degrees). The vertical measurements, similarly marked, give the altitudes of the stars up to $62\frac{1}{2}$ degrees. Estimates of altitude and azimuth made from these charts will necessarily be mere approximations, since no observer will be exactly at the adopted latitude, or at the stated time, but they will serve for the identification of stars and planets.

The ecliptic is drawn as a broken line on which longitude is marked at every 10 degrees; the positions of the planets are then easily found by reference to the table on page 71. It will be noticed that on the Southern Charts the **ecliptic** may reach an altitude in excess of 62½ degrees on star charts 5 to 9. The continuations of the broken line will be found on the charts of overhead stars.

There is a curious illusion that stars at an altitude of 60 degrees or more are actually overhead, and the beginner may often feel that he is leaning over backwards in trying to see them. These overhead stars are given separately on the pages immediately following the main star charts. The entire year is covered at one opening, each of the four maps showing the overhead stars at times which correspond to those of three of the main star charts. The position of the zenith is indicated by a cross, and this cross marks the centre of a circle which is 35 degrees from the zenith; there is thus a small overlap with the main charts.

The broken line leading from the north (on the Northern Charts) or from the south (on the Southern Charts) is numbered to indicate the corresponding main chart. Thus on page 38 the N-S line numbered 6 is to be regarded as an extension of the centre (south) line of chart 6 on pages 24 and 25, and at the top of these pages are printed the dates and times which are appropriate. Similarly, on page 65, the S-N line numbered 10 connects with the north line of the upper charts on pages 58 and 59.

The overhead stars are plotted as maps on a conical projection, and the scale is rather smaller than that of the main charts.

1L

October 6 at 5ʰ   October 21 at 4ʰ
November 6 at 3ʰ   November 21 at 2ʰ
December 6 at 1ʰ   December 21 at midnight
January 6 at 23ʰ   January 21 at 22ʰ
February 6 at 21ʰ   February 21 at 20ʰ

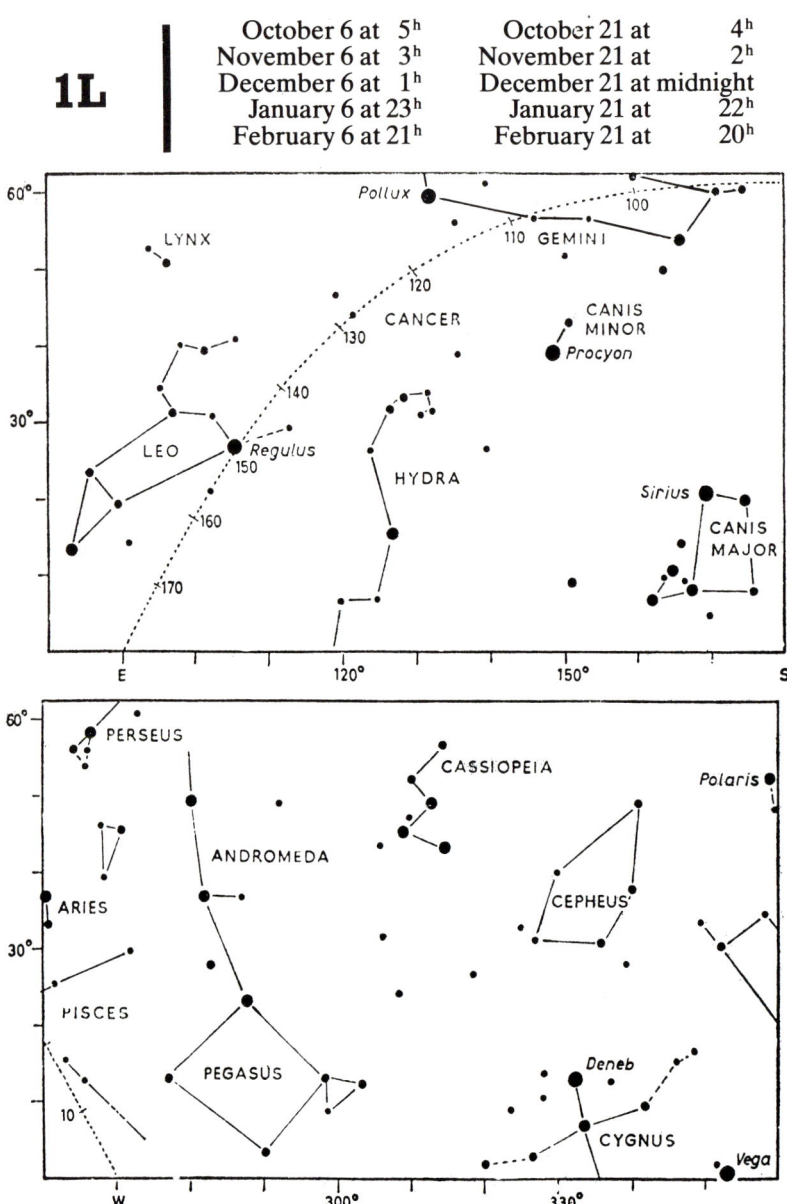

October 6 at 5ʰ   October 21 at   4ʰ
November 6 at 3ʰ   November 21 at   2ʰ
December 6 at 1ʰ   December 21 at midnight
January 6 at 23ʰ   January 21 at   22ʰ
February 6 at 21ʰ   February 21 at   20ʰ

1R

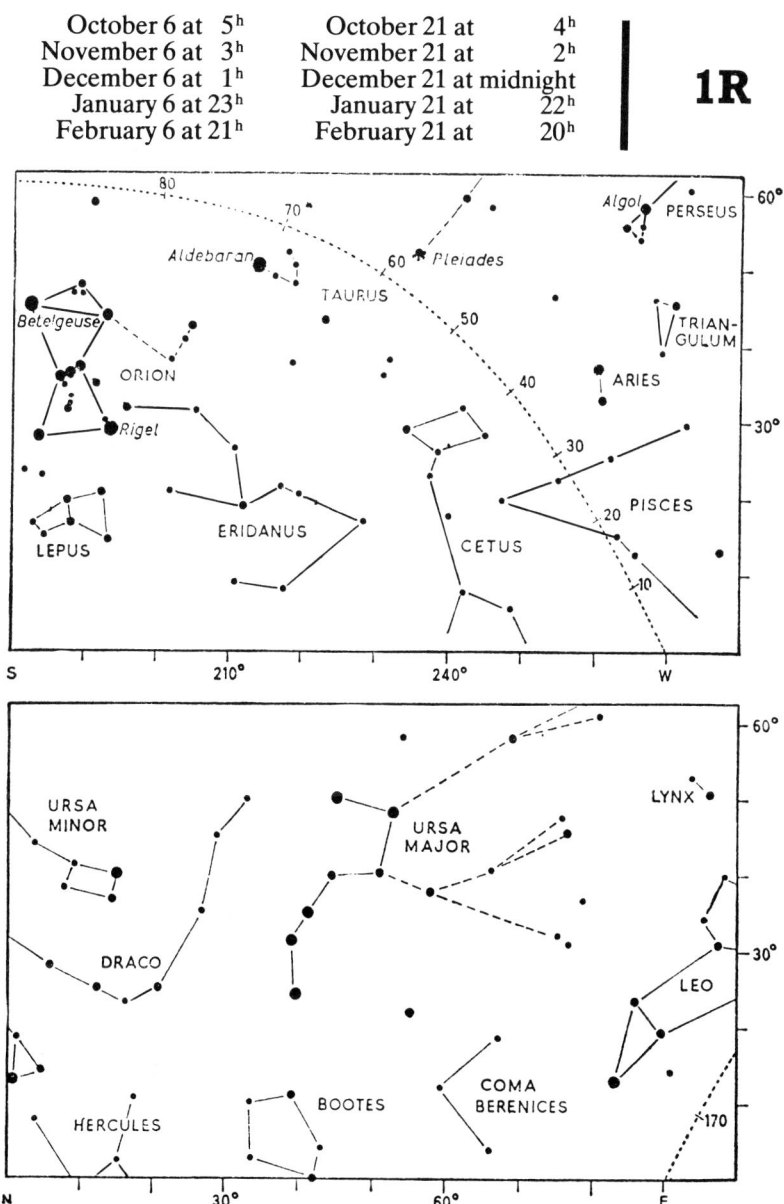

2L

November 6 at 5^h	November 21 at 4^h
December 6 at 3^h	December 21 at 2^h
January 6 at 1^h	January 21 at midnight
February 6 at 23^h	February 21 at 22^h
March 6 at 21^h	March 21 at 20^h

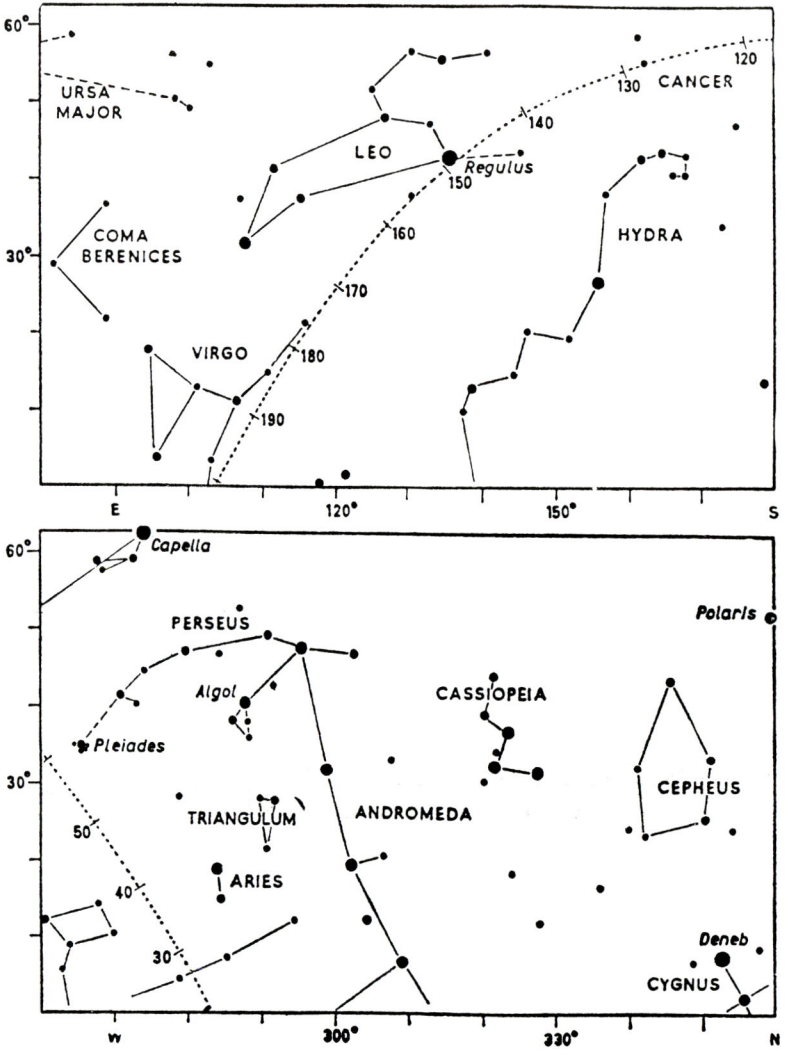

November 6 at 5ʰ	November 21 at 4ʰ
December 6 at 3ʰ	December 21 at 2ʰ
January 6 at 1ʰ	January 21 at midnight
February 6 at 23ʰ	February 21 at 22ʰ
March 6 at 21ʰ	March 21 at 20ʰ

2R

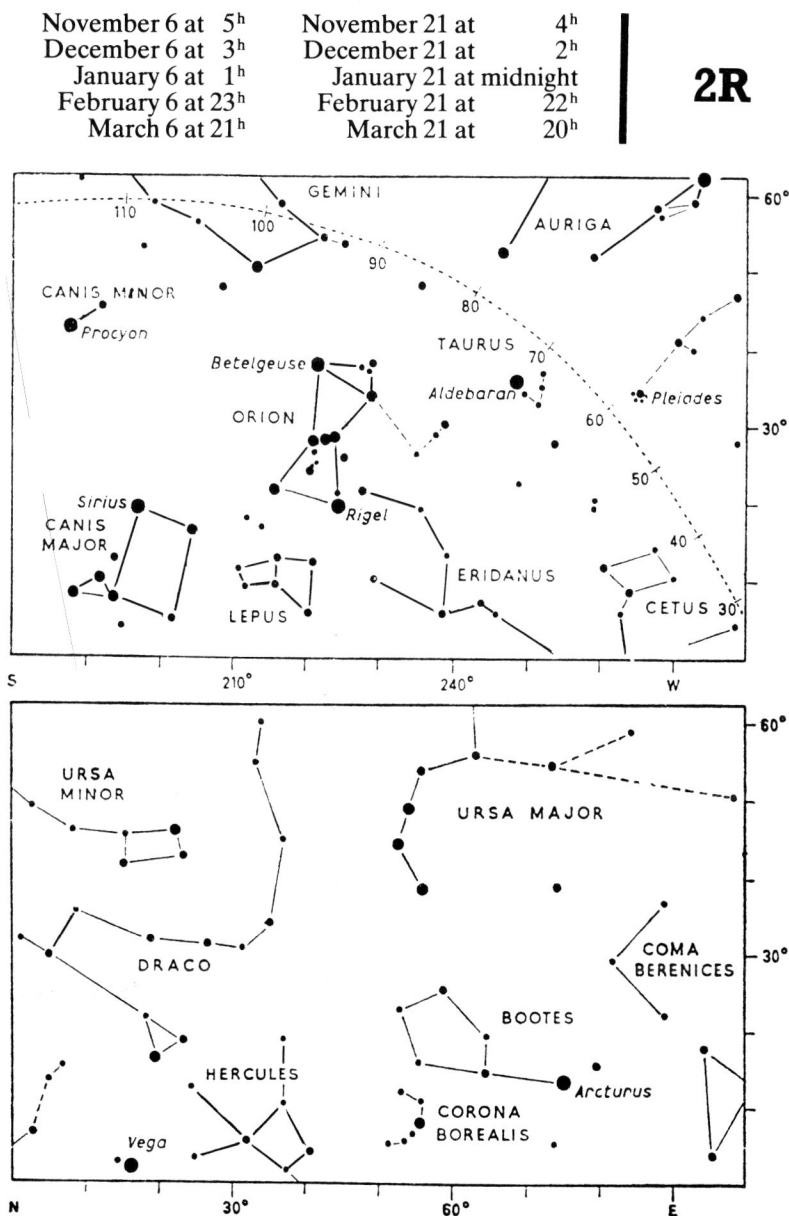

17

3L

December 6 at 5ʰ	December 21 at 4ʰ
January 6 at 3ʰ	January 21 at 2ʰ
February 6 at 1ʰ	February 21 at midnight
March 6 at 23ʰ	March 21 at 22ʰ
April 6 at 21ʰ	April 21 at 20ʰ

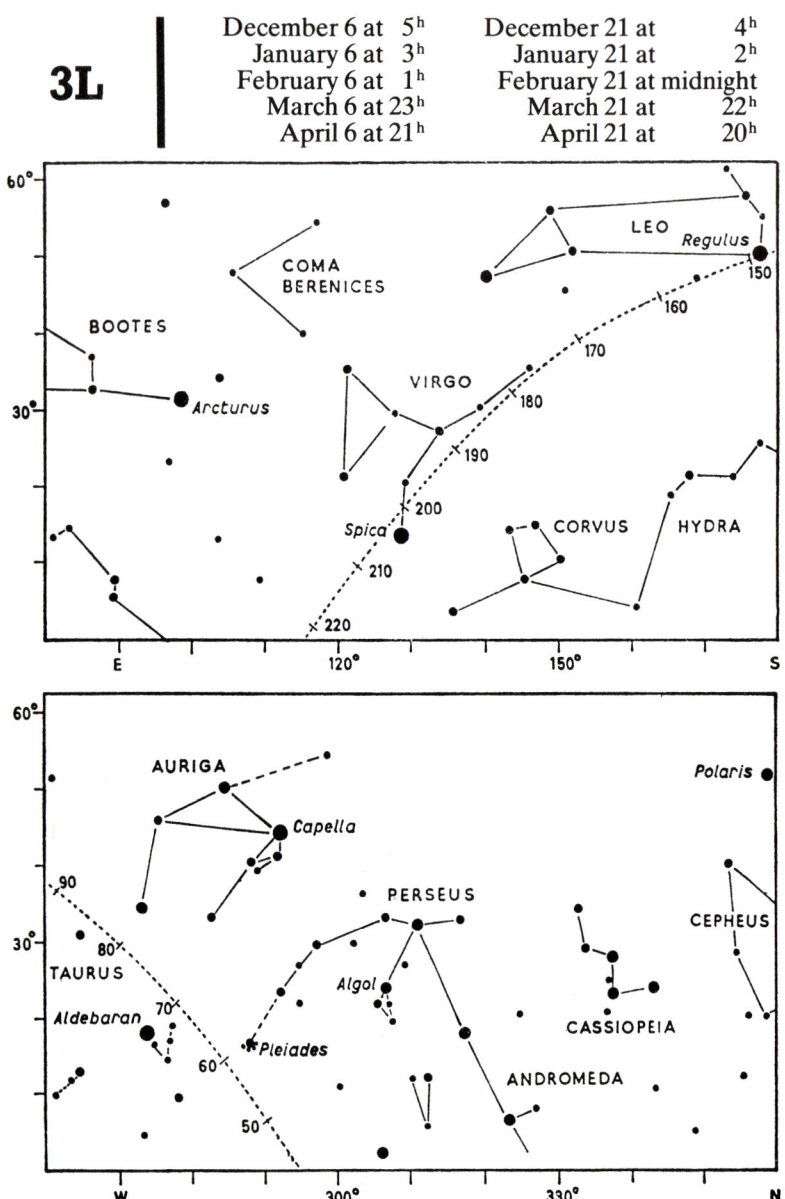

December 6 at 5ʰ	December 21 at 4ʰ
January 6 at 3ʰ	January 21 at 2ʰ
February 6 at 1ʰ	February 21 at midnight
March 6 at 23ʰ	March 21 at 22ʰ
April 6 at 21ʰ	April 21 at 20ʰ

3R

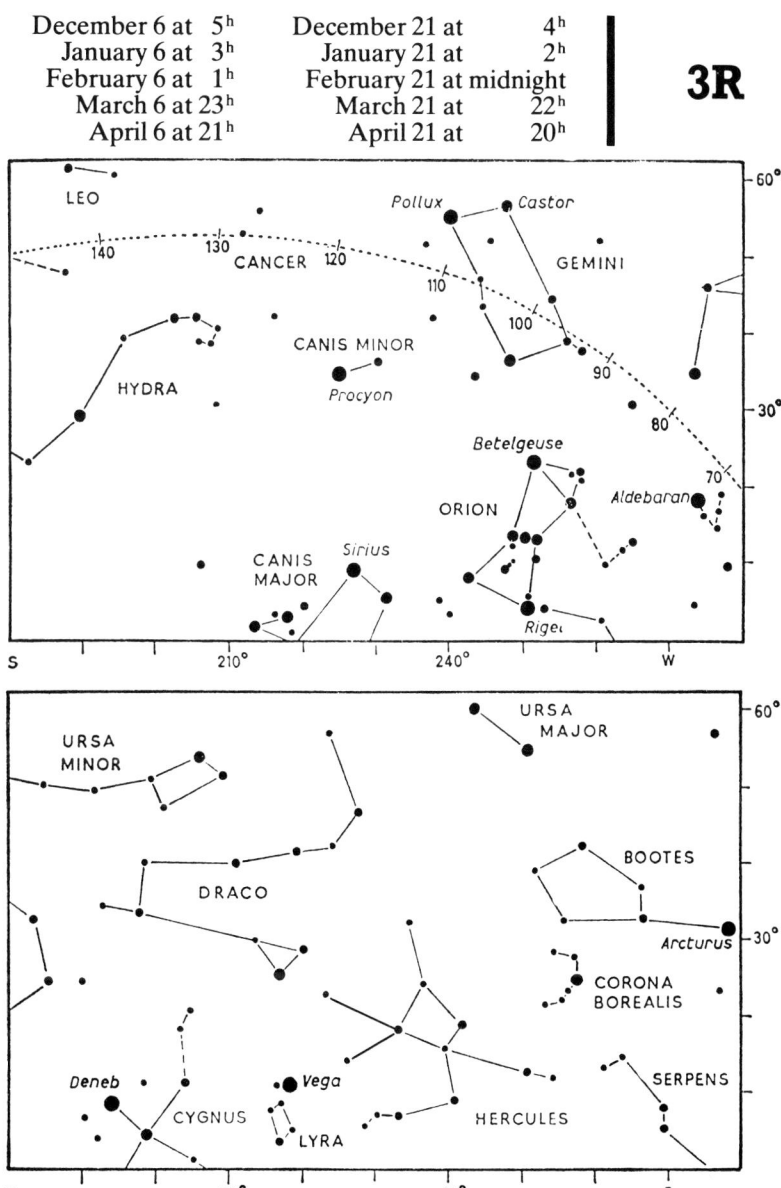

4L

January 6 at 5^h
February 6 at 3^h
March 6 at 1^h
April 6 at 23^h
May 6 at 21^h

January 21 at 4^h
February 21 at 2^h
March 21 at midnight
April 21 at 22^h
May 21 at 20^h

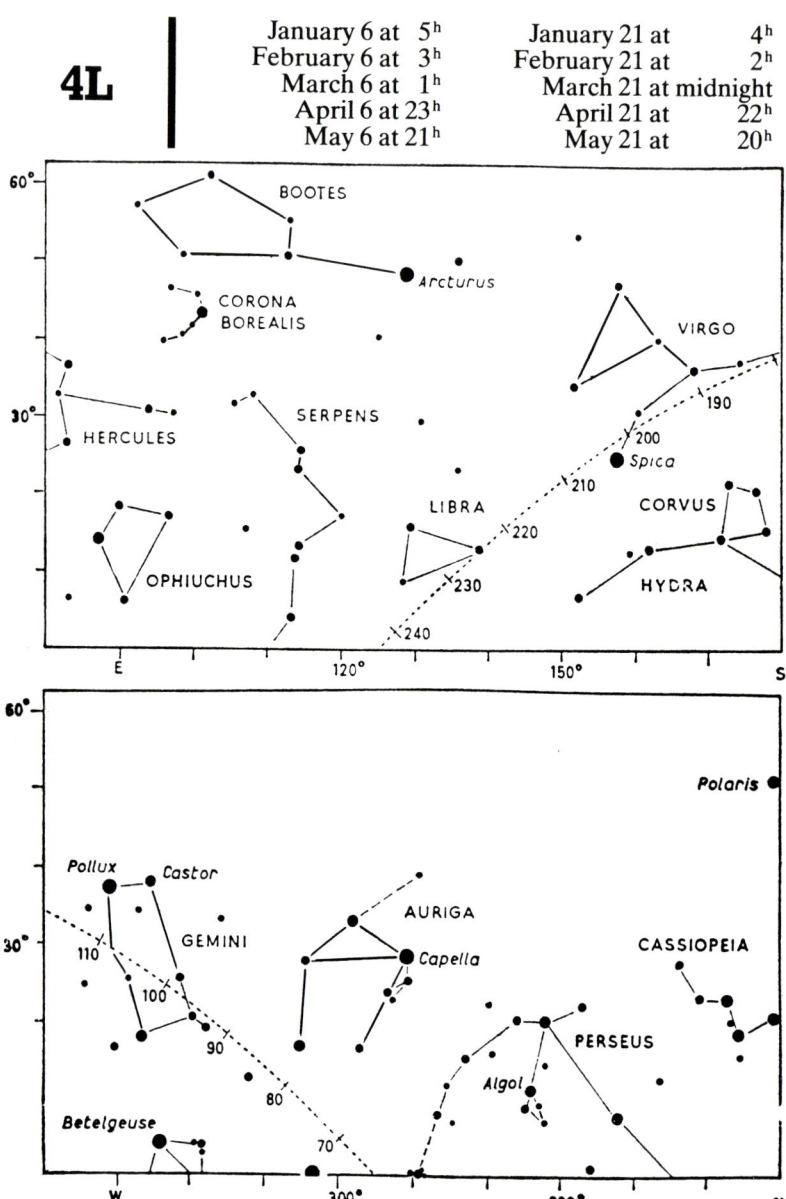

January 6 at 5^h January 21 at   4^h
February 6 at 3^h February 21 at   2^h
March 6 at 1^h March 21 at midnight
April 6 at 23^h April 21 at   22^h
May 6 at 21^h May 21 at   20^h

4R

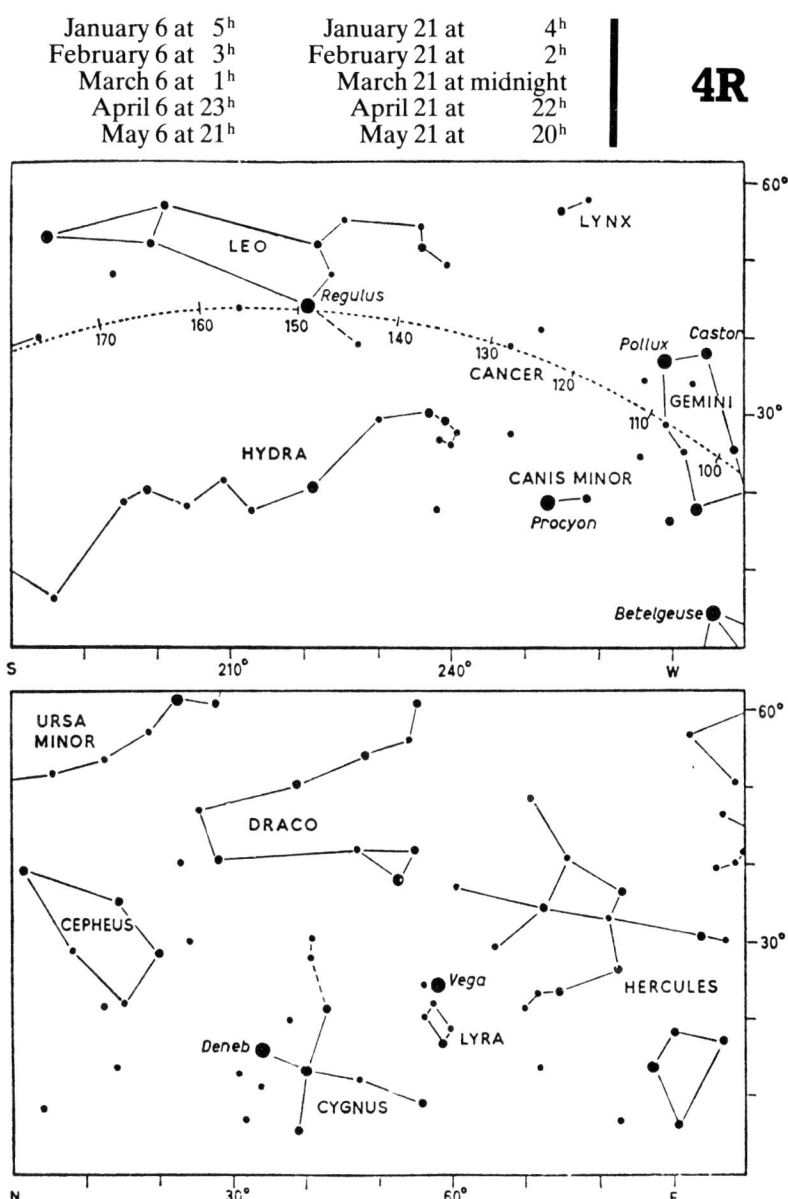

21

5L

January 6 at 7ʰ   January 21 at 6ʰ
February 6 at 5ʰ   February 21 at 4ʰ
March 6 at 3ʰ   March 21 at 2ʰ
April 6 at 1ʰ   April 21 at midnight
May 6 at 23ʰ   May 21 at 22ʰ

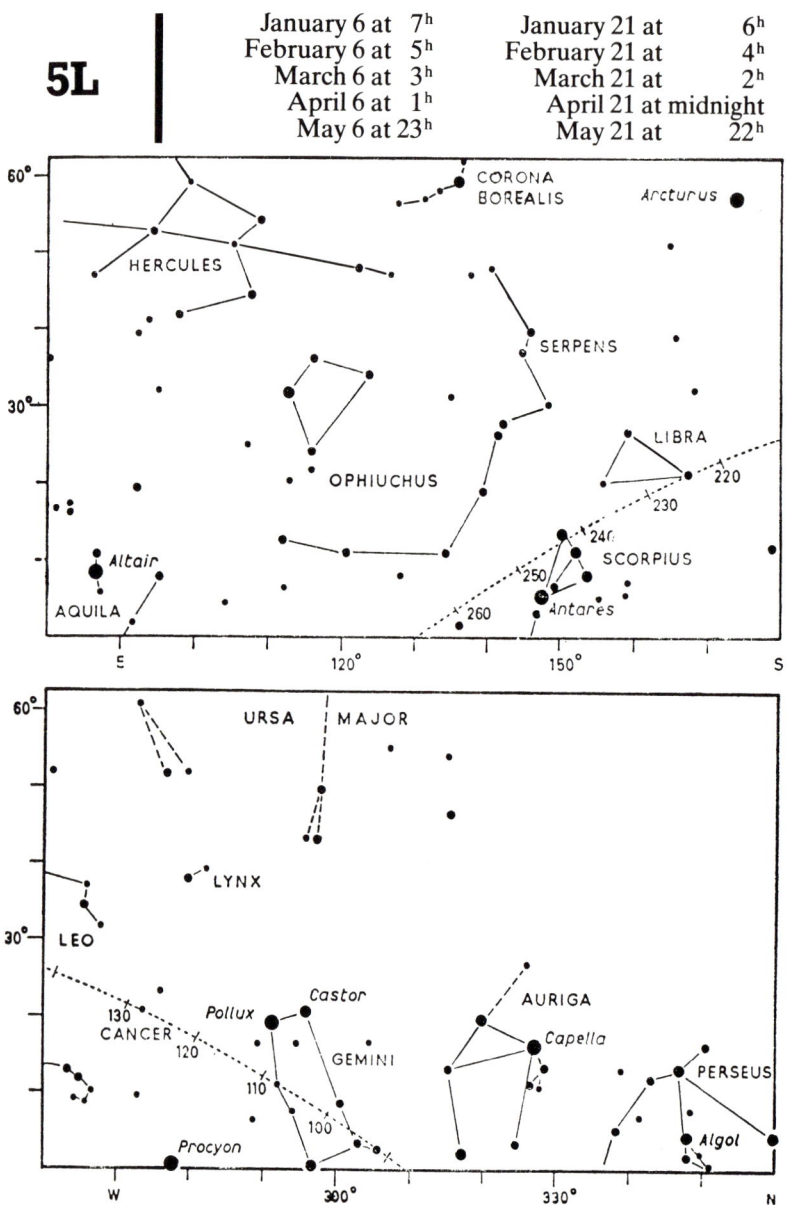

January 6 at 7ʰ January 21 at   6ʰ
February 6 at 5ʰ February 21 at   4ʰ
March 6 at 3ʰ March 21 at   2ʰ
April 6 at 1ʰ April 21 at midnight
May 6 at 23ʰ May 21 at   22ʰ

5R

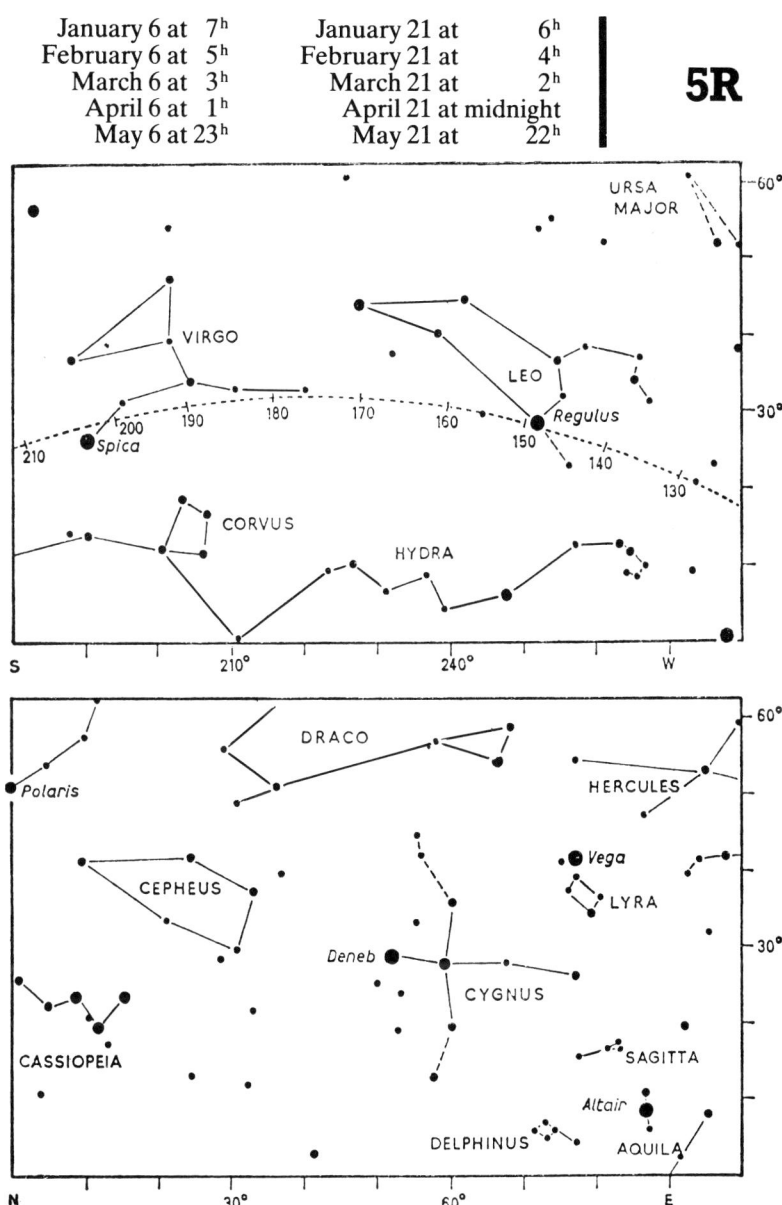

23

6L

March 6 at 5^h

April 6 at 3^h

May 6 at 1^h

June 6 at 23^h

July 6 at 21^h

March 21 at 4^h

April 21 at 2^h

May 21 at midnight

June 21 at 22^h

July 21 at 20^h

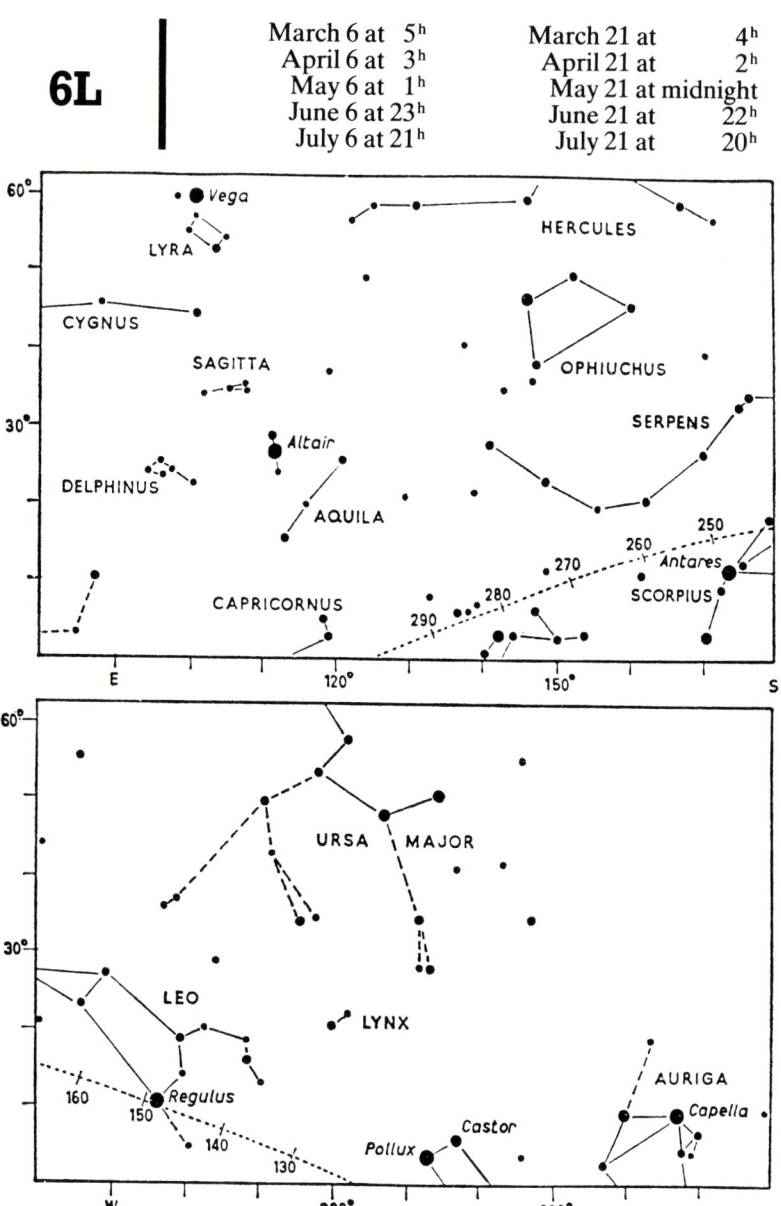

March 6 at 5^h March 21 at 4^h
April 6 at 3^h April 21 at 2^h
May 6 at 1^h May 21 at midnight
June 6 at 23^h June 21 at 22^h
July 6 at 21^h July 21 at 20^h

6R

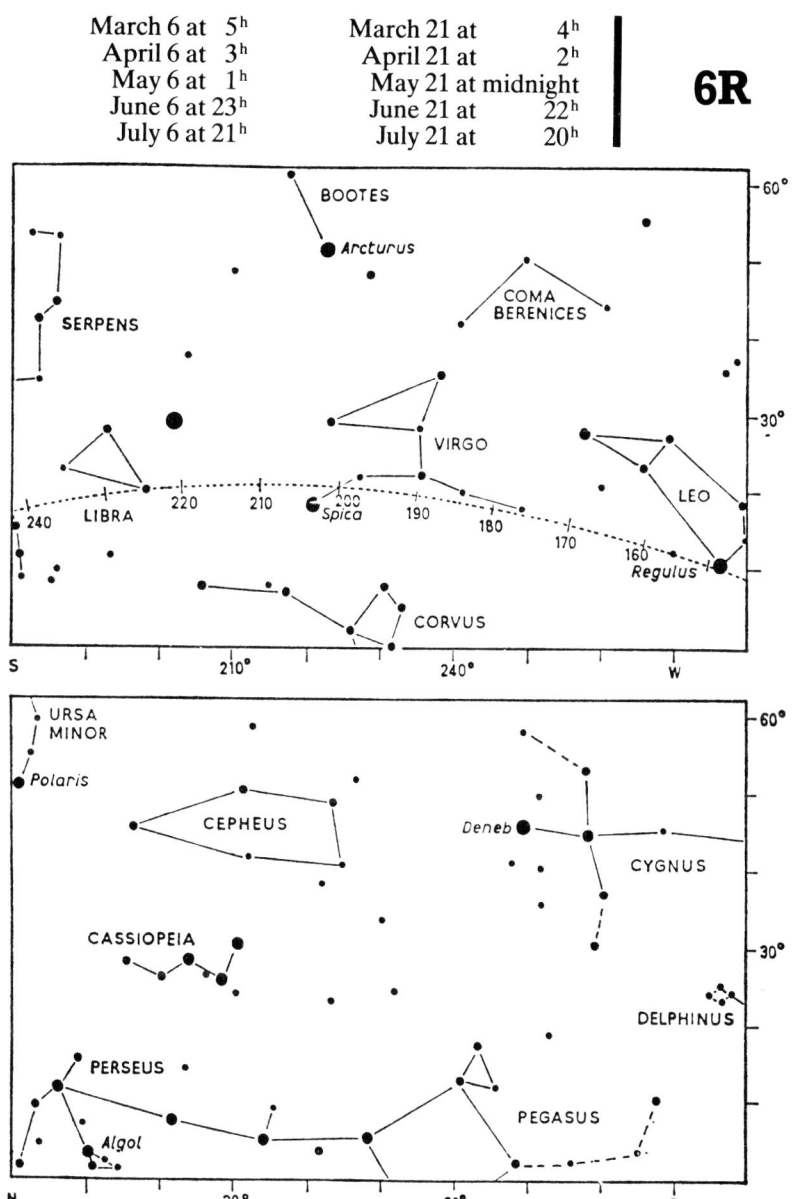

25

7L

May 6 at 3ʰ	May 21 at 2ʰ
June 6 at 1ʰ	June 21 at midnight
July 6 at 23ʰ	July 21 at 22ʰ
August 6 at 21ʰ	August 21 at 20ʰ
September 6 at 19ʰ	September 21 at 18ʰ

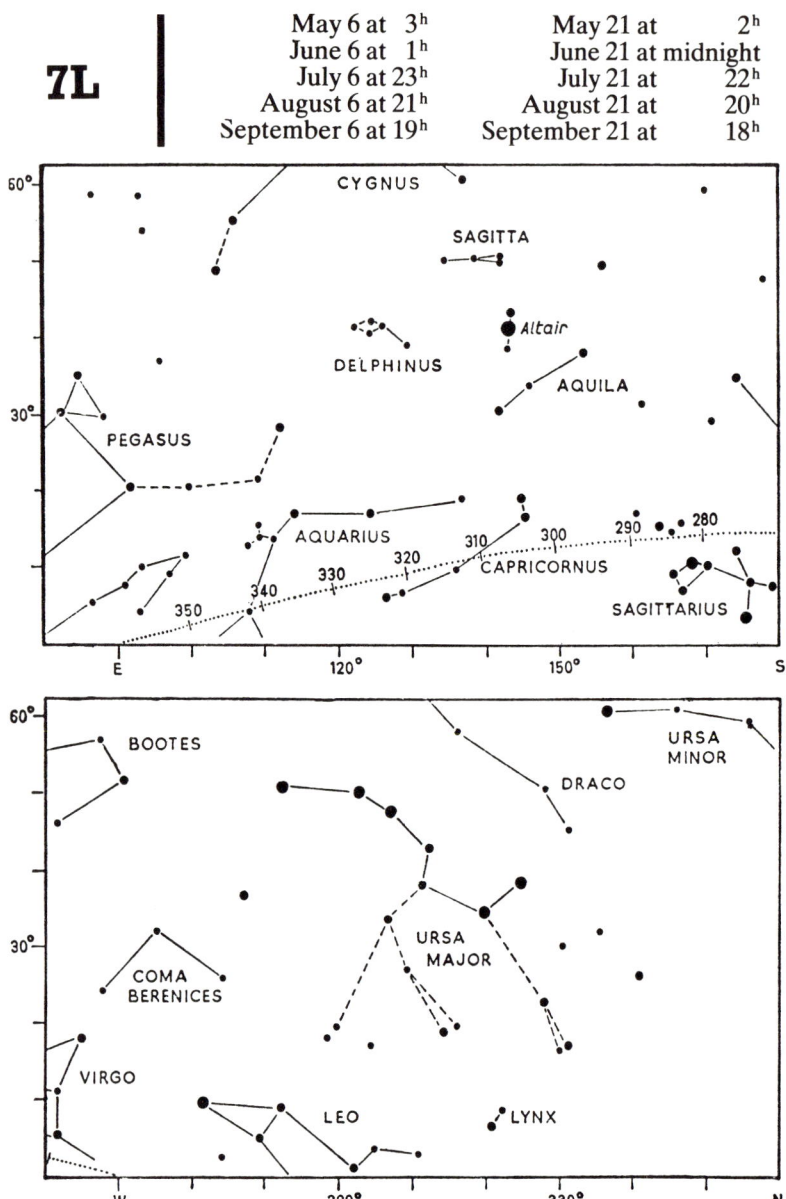

May 6 at   3ʰ   May 21 at   2ʰ
June 6 at   1ʰ   June 21 at midnight
July 6 at 23ʰ   July 21 at   22ʰ
August 6 at 21ʰ   August 21 at   20ʰ
September 6 at 19ʰ   September 21 at   18ʰ

7R

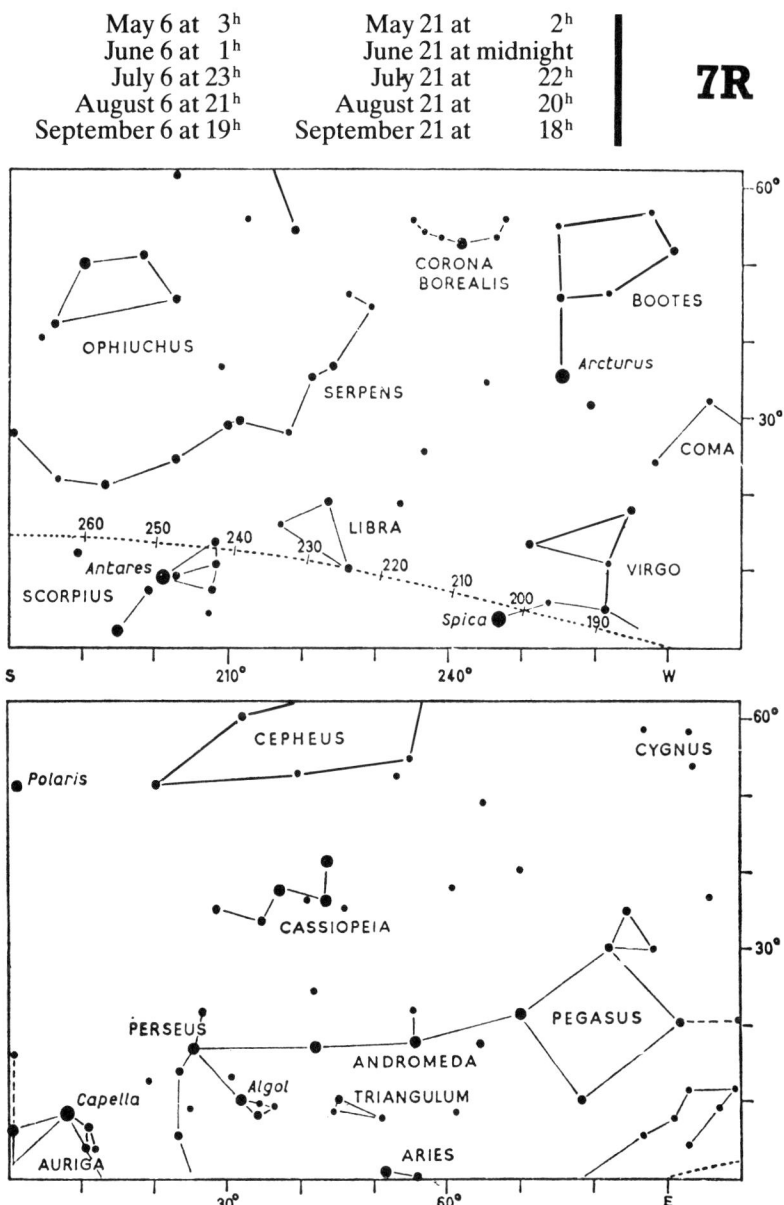

8L

July 6 at 1ʰ	July 21 at midnight
August 6 at 23ʰ	August 21 at 22ʰ
September 6 at 21ʰ	September 21 at 20ʰ
October 6 at 19ʰ	October 21 at 18ʰ
November 6 at 17ʰ	November 21 at 16ʰ

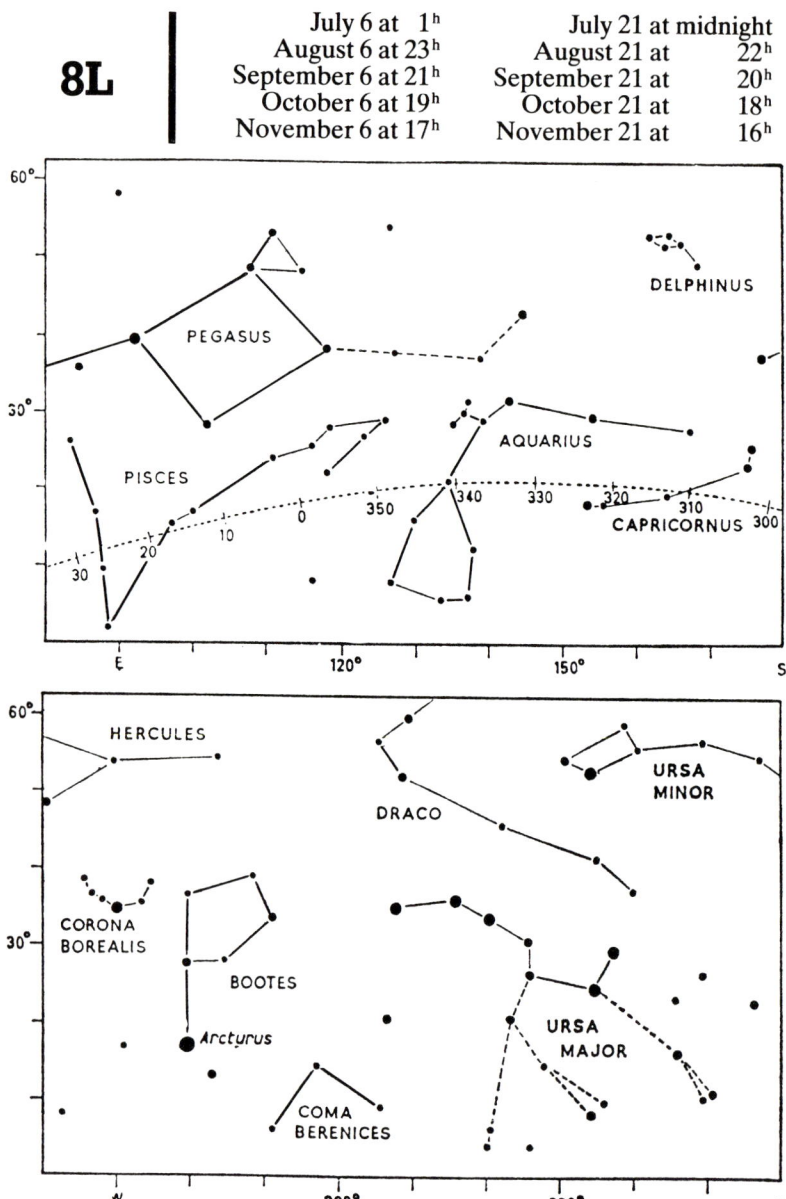

July 6 at   1ʰ   July 21 at midnight
August 6 at 23ʰ   August 21 at   22ʰ
September 6 at 21ʰ   September 21 at   20ʰ   **8R**
October 6 at 19ʰ   October 21 at   18ʰ
November 6 at 17ʰ   November 21 at   16ʰ

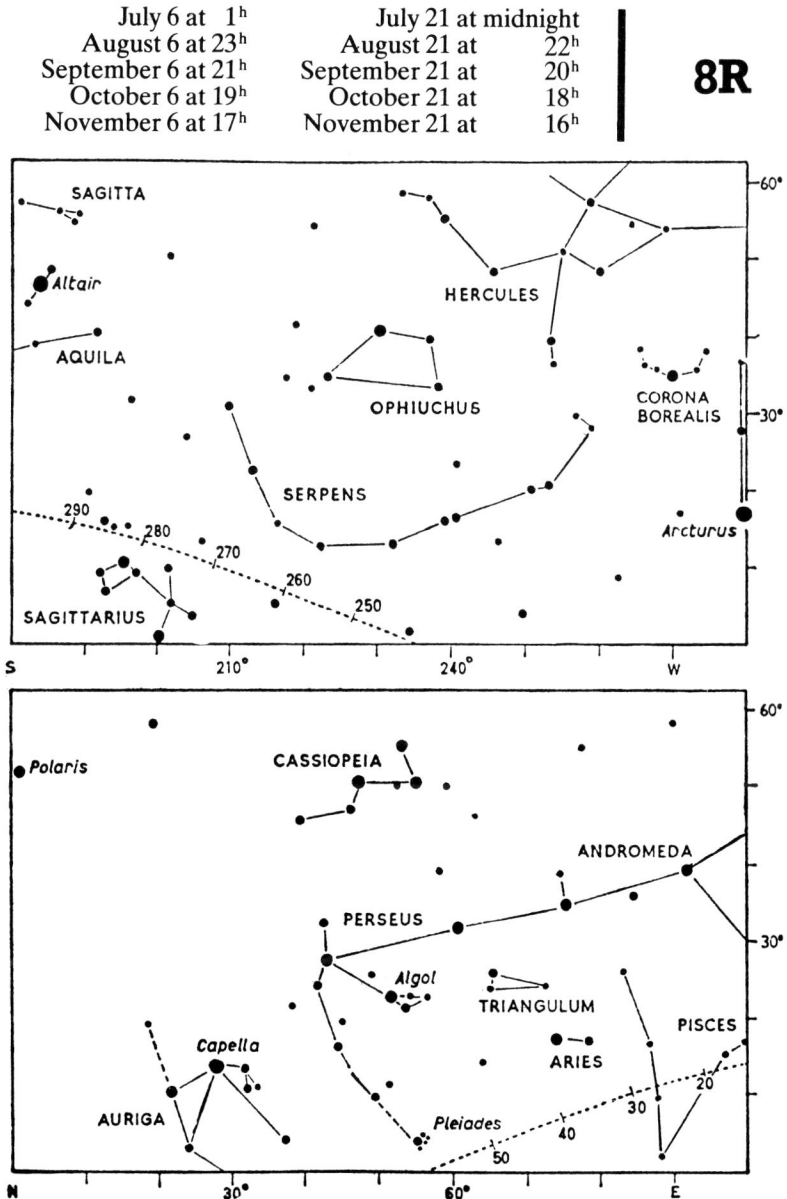

9L

August 6 at 1^h	August 21 at midnight
September 6 at 23^h	September 21 at 22^h
October 6 at 21^h	October 21 at 20^h
November 6 at 19^h	November 21 at 18^h
December 6 at 17^h	December 21 at 16^h

August 6 at 1ʰ   August 21 at midnight
September 6 at 23ʰ   September 21 at   22ʰ
October 6 at 21ʰ   October 21 at   20ʰ
November 6 at 19ʰ   November 21 at   18ʰ
December 6 at 17ʰ   December 21 at   16ʰ

9R

10L

August 6 at 3^h	August 21 at 2^h
September 6 at 1^h	September 21 at midnight
October 6 at 23^h	October 21 at 22^h
November 6 at 21^h	November 21 at 20^h
December 6 at 19^h	December 21 at 18^h

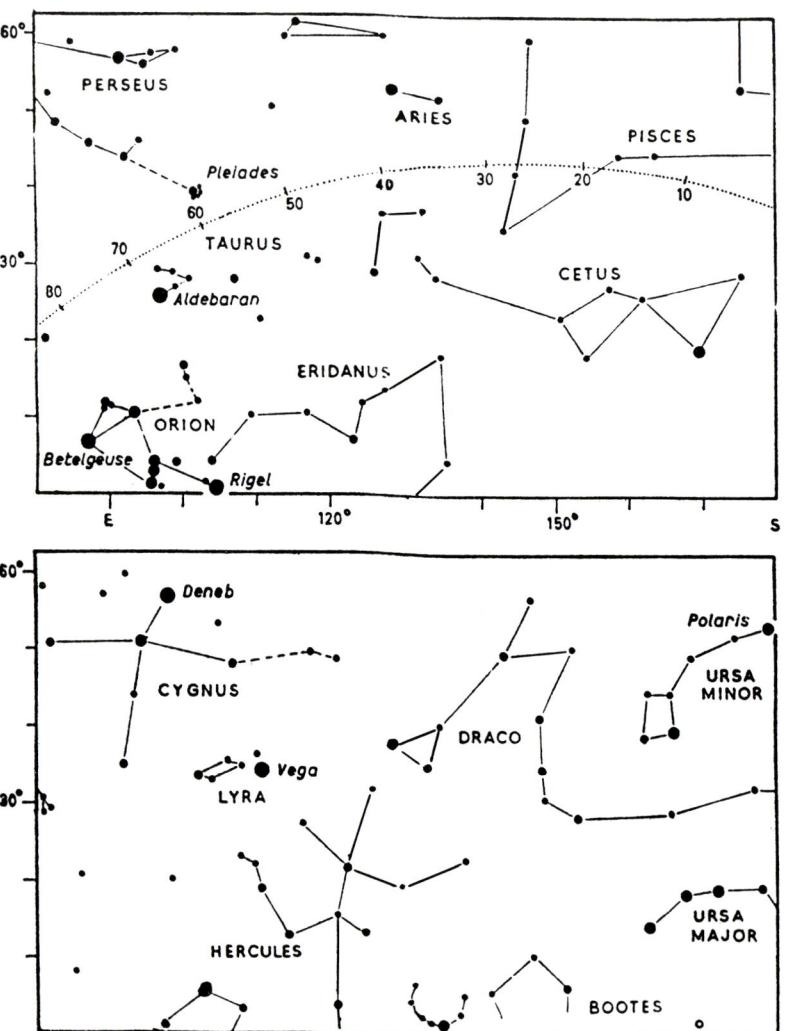

August 6 at 3ʰ	August 21 at 2ʰ	
September 6 at 1ʰ	September 21 at midnight	
October 6 at 23ʰ	October 21 at 22ʰ	**10R**
November 6 at 21ʰ	November 21 at 20ʰ	
December 6 at 19ʰ	December 21 at 18ʰ	

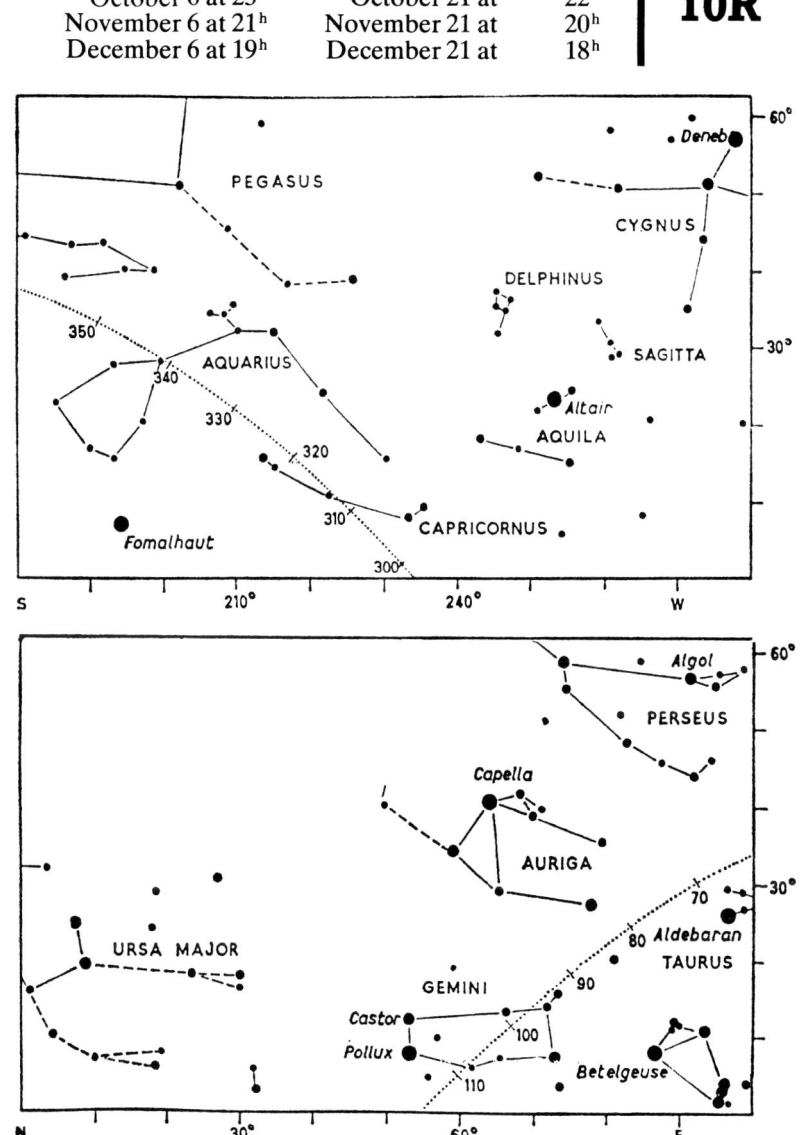

33

11L

September 6 at 3ʰ September 21 at 2ʰ
October 6 at 1ʰ October 21 at midnight
November 6 at 23ʰ November 21 at 22ʰ
December 6 at 21ʰ December 21 at 20ʰ
January 6 at 19ʰ January 21 at 18ʰ

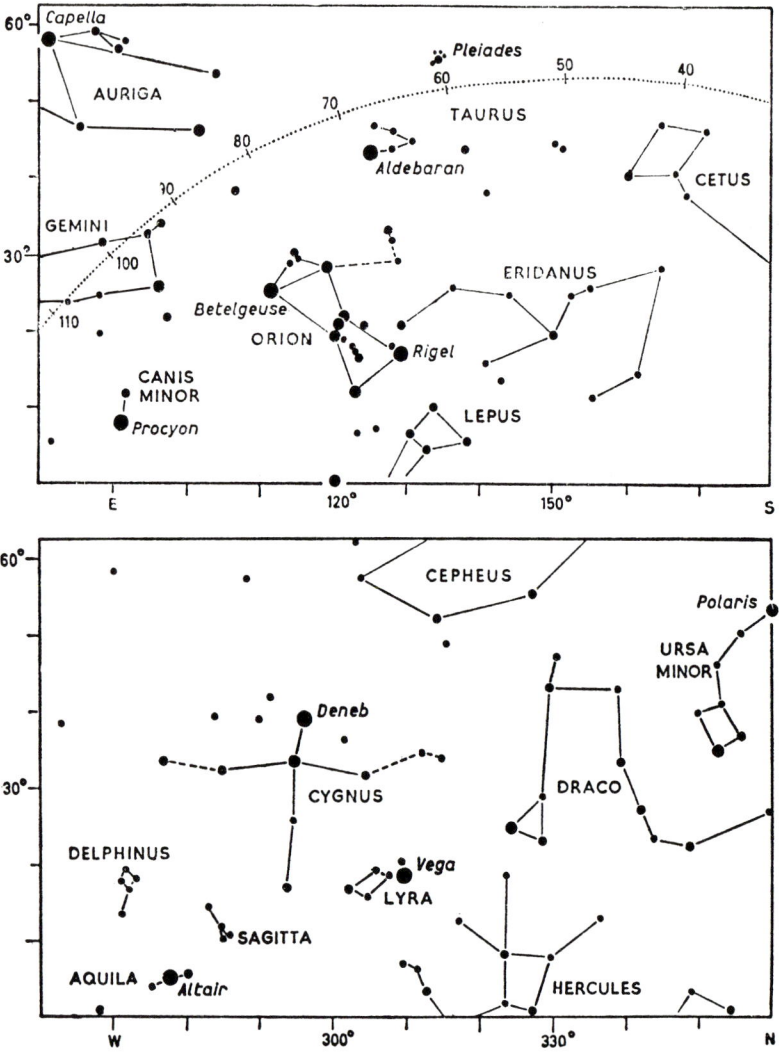

September 6 at 3ʰ September 21 at 2ʰ
October 6 at 1ʰ October 21 at midnight
November 6 at 23ʰ November 21 at 22ʰ
December 6 at 21ʰ December 21 at 20ʰ
January 6 at 19ʰ January 21 at 18ʰ

11R

12L

October 6 at 3ʰ	October 21 at 2ʰ
November 6 at 1ʰ	November 21 at midnight
December 6 at 23ʰ	December 21 at 22ʰ
January 6 at 21ʰ	January 21 at 20ʰ
February 6 at 19ʰ	February 21 at 18ʰ

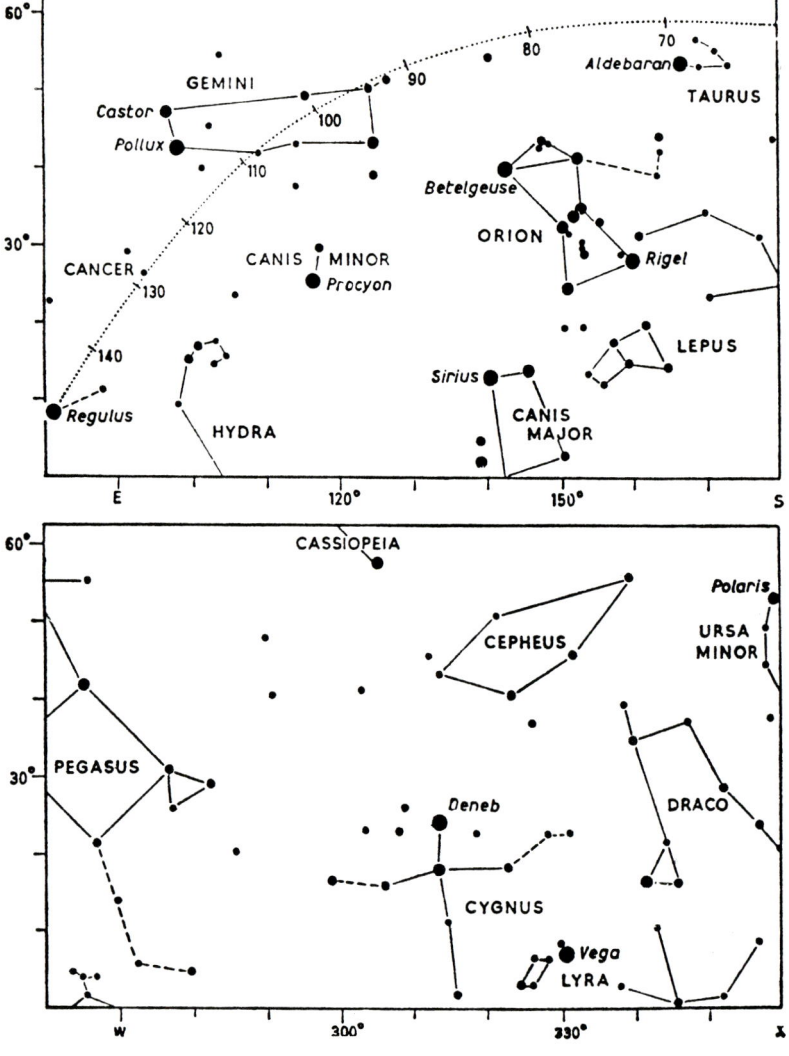

October 6 at 3^h   October 21 at   2^h
November 6 at 1^h   November 21 at midnight
December 6 at 23^h   December 21 at   22^h
January 6 at 21^h   January 21 at   20^h
February 6 at 19^h   February 21 at   18^h

12R

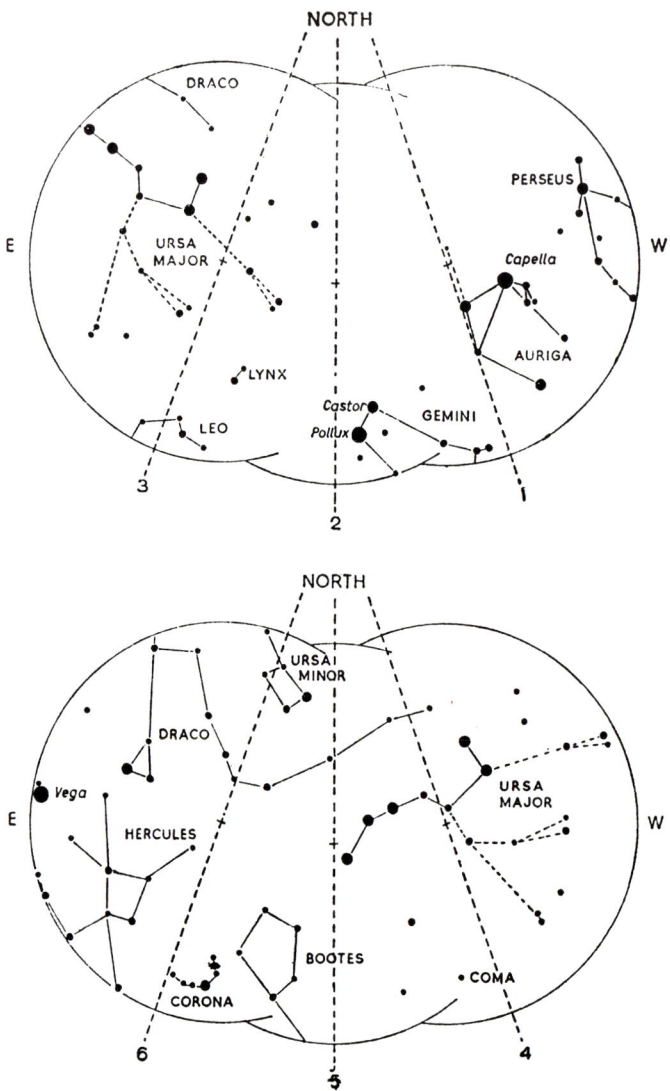

Northern Hemisphere Overhead Stars

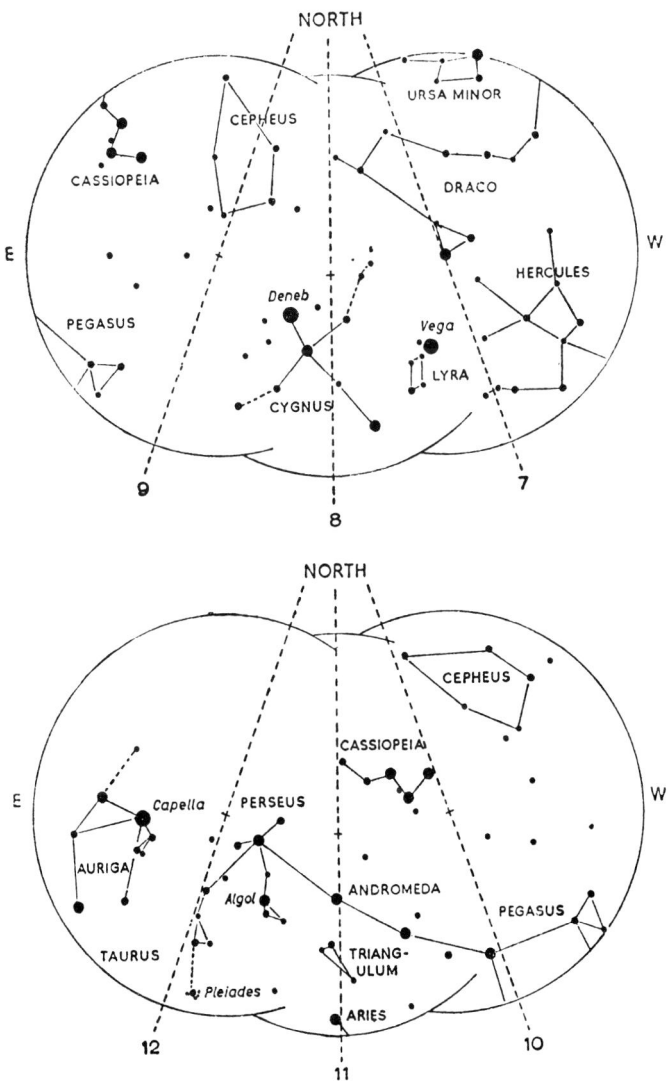

Northern Hemisphere Overhead Stars

1L

October 6 at 5^h	October 21 at 4^h
November 6 at 3^h	November 21 at 2^h
December 6 at 1^h	December 21 at midnight
January 6 at 23^h	January 21 at 22^h
February 6 at 21^h	February 21 at 20^h

October 6 at 5^h   October 21 at   4^h
November 6 at 3^h   November 21 at   2^h
December 6 at 1^h   December 21 at midnight
January 6 at 23^h   January 21 at   22^h
February 6 at 21^h   February 21 at   20^h

1R

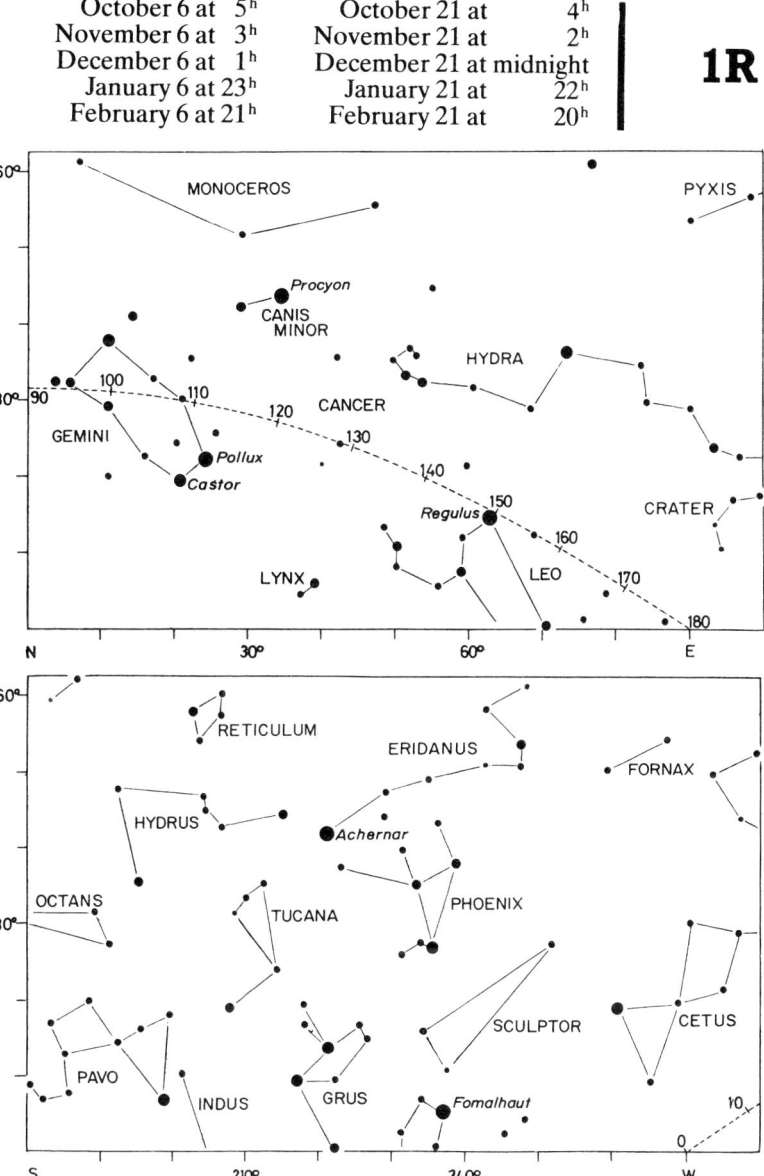

2L

November 6 at 5ʰ	November 21 at 4ʰ
December 6 at 3ʰ	December 21 at 2ʰ
January 6 at 1ʰ	January 21 at midnight
February 6 at 23ʰ	February 21 at 22ʰ
March 6 at 21ʰ	March 21 at 20ʰ

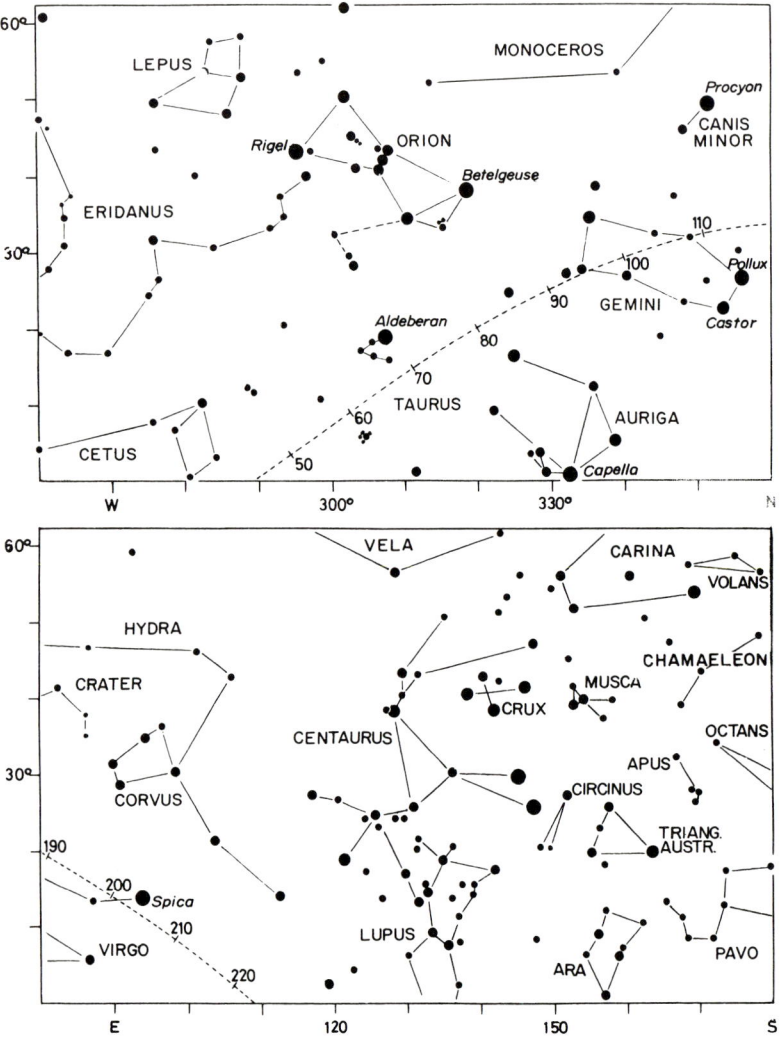

November 6 at 5ʰ	November 21 at	4ʰ
December 6 at 3ʰ	December 21 at	2ʰ
January 6 at 1ʰ	January 21 at midnight	
February 6 at 23ʰ	February 21 at	22ʰ
March 6 at 21ʰ	March 21 at	20ʰ

2R

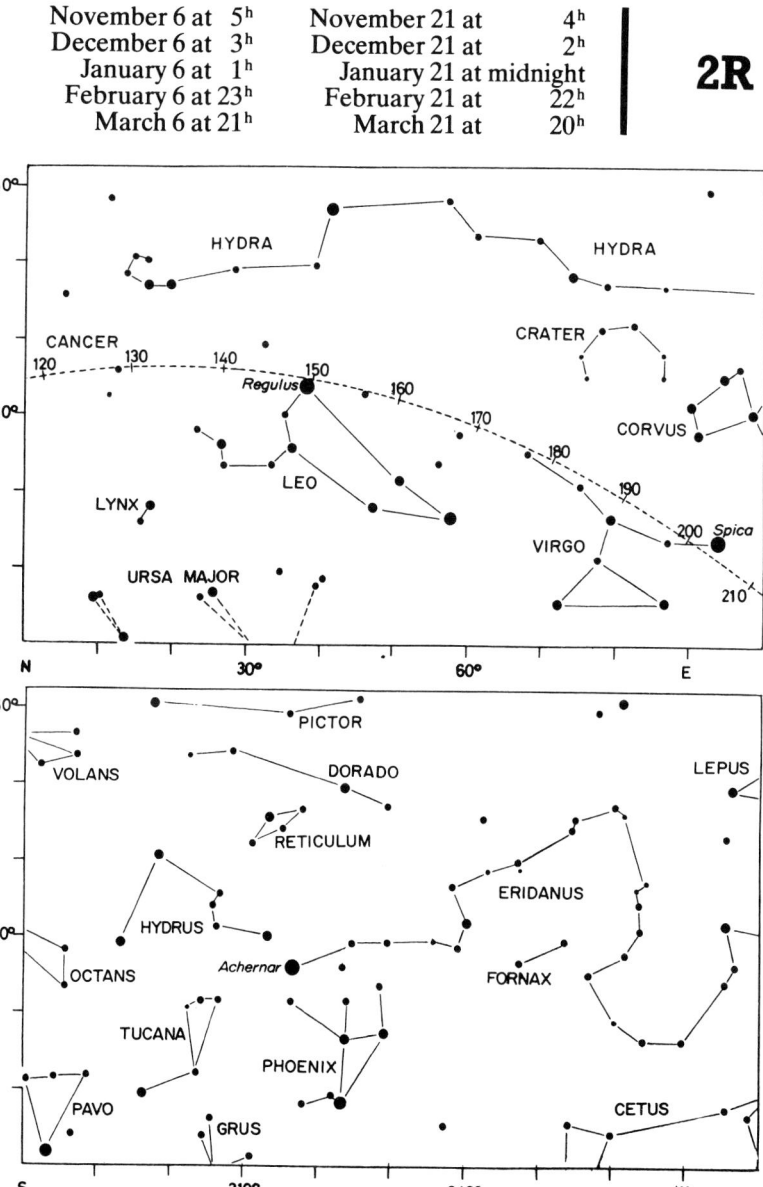

3L

January 6 at 3ʰ	January 21 at 2ʰ
February 6 at 1ʰ	February 21 at midnight
March 6 at 23ʰ	March 21 at 22ʰ
April 6 at 21ʰ	April 21 at 20ʰ
May 6 at 19ʰ	May 21 at 18ʰ

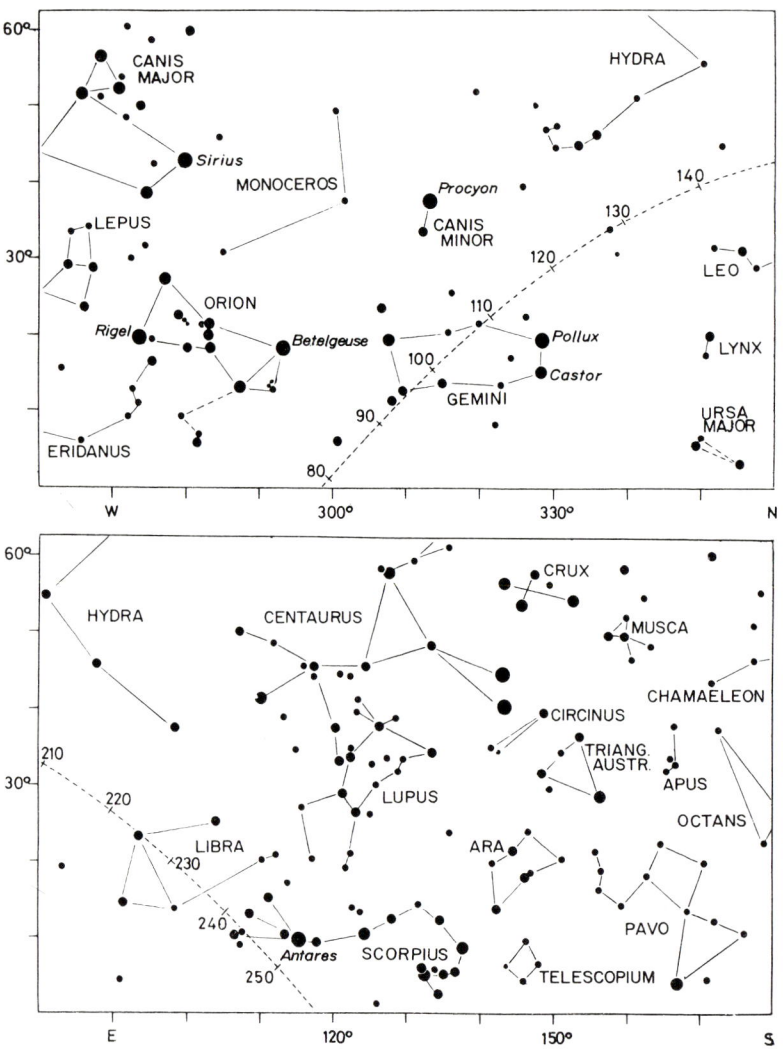

January 6 at 3ʰ	January 21 at 2ʰ	
February 6 at 1ʰ	February 21 at midnight	**3R**
March 6 at 23ʰ	March 21 at 22ʰ	
April 6 at 21ʰ	April 21 at 20ʰ	
May 6 at 19ʰ	May 21 at 18ʰ	

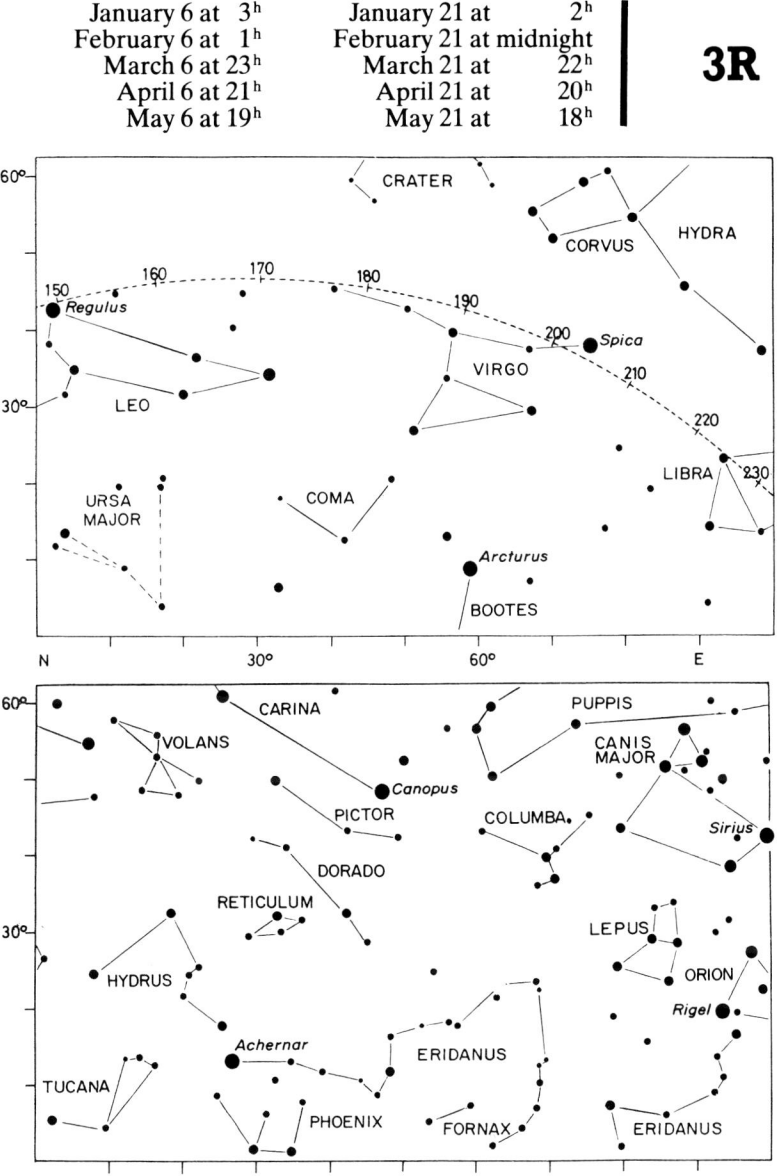

4L

February 6 at 3ʰ	February 21 at 2ʰ
March 6 at 1ʰ	March 21 at midnight
April 6 at 23ʰ	April 21 at 22ʰ
May 6 at 21ʰ	May 21 at 20ʰ
June 6 at 19ʰ	June 21 at 18ʰ

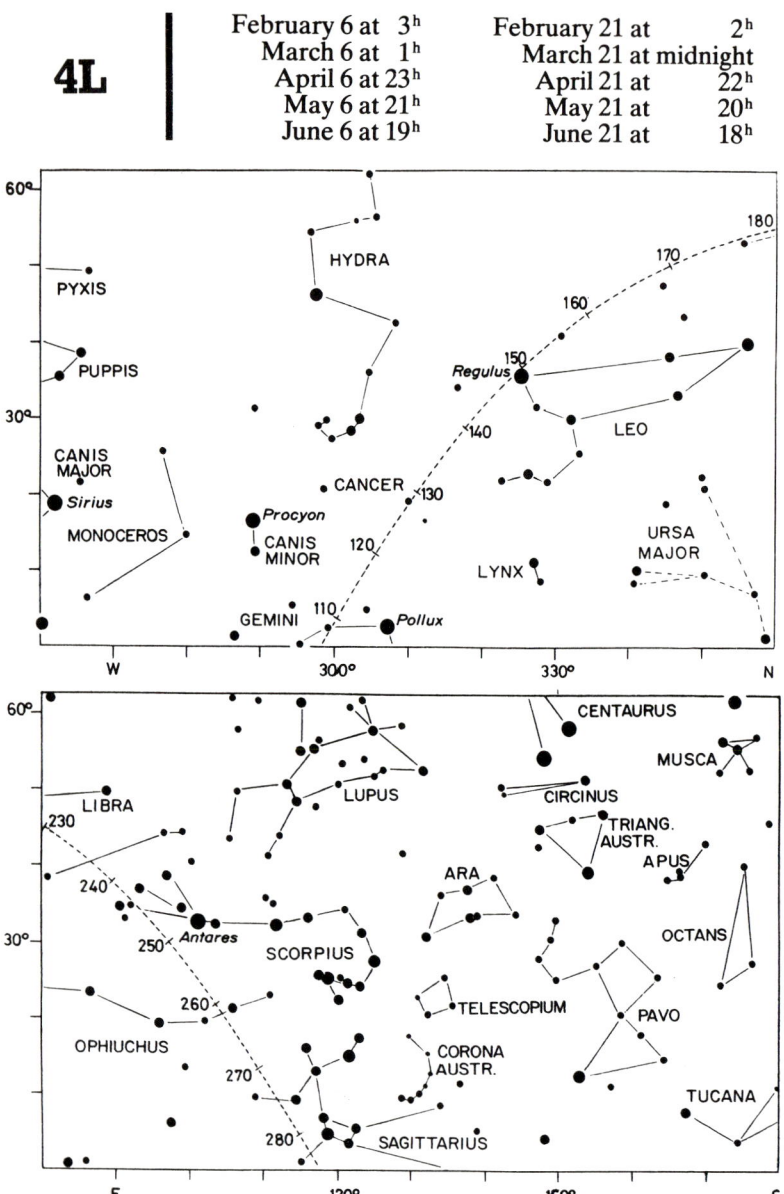

February 6 at 3ʰ	February 21 at 2ʰ	
March 6 at 1ʰ	March 21 at midnight	
April 6 at 23ʰ	April 21 at 22ʰ	**4R**
May 6 at 21ʰ	May 21 at 20ʰ	
June 6 at 19ʰ	June 21 at 18ʰ	

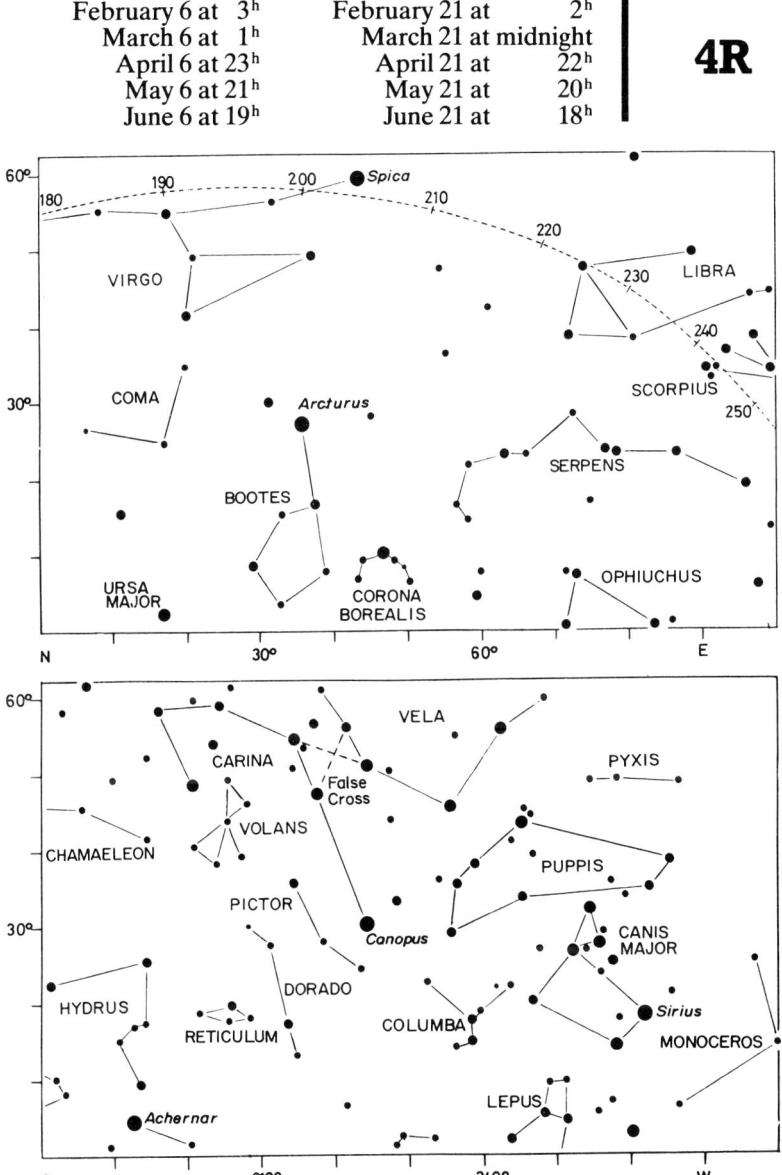

47

5L

March 6 at 3ʰ	March 21 at 2ʰ
April 6 at 1ʰ	April 21 at midnight
May 6 at 23ʰ	May 21 at 22ʰ
June 6 at 21ʰ	June 21 at 20ʰ
July 6 at 19ʰ	July 21 at 18ʰ

March 6 at 3^h	March 21 at 2^h
April 6 at 1^h	April 21 at midnight
May 6 at 23^h	May 21 at 22^h
June 6 at 21^h	June 21 at 20^h
July 6 at 19^h	July 21 at 18^h

5R

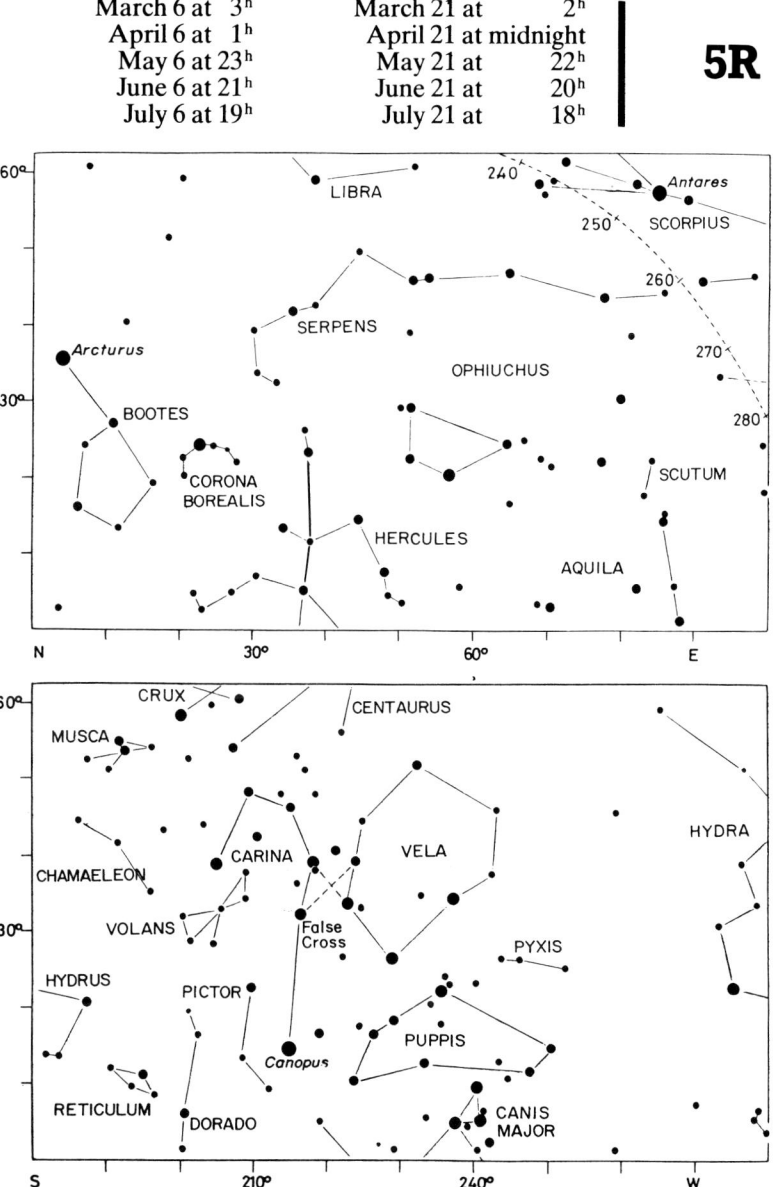

6L

March 6 at 5^h March 21 at   4^h
April 6 at 3^h April 21 at   2^h
May 6 at 1^h May 21 at midnight
June 6 at 23^h June 21 at   22^h
July 6 at 21^h July 21 at   20^h

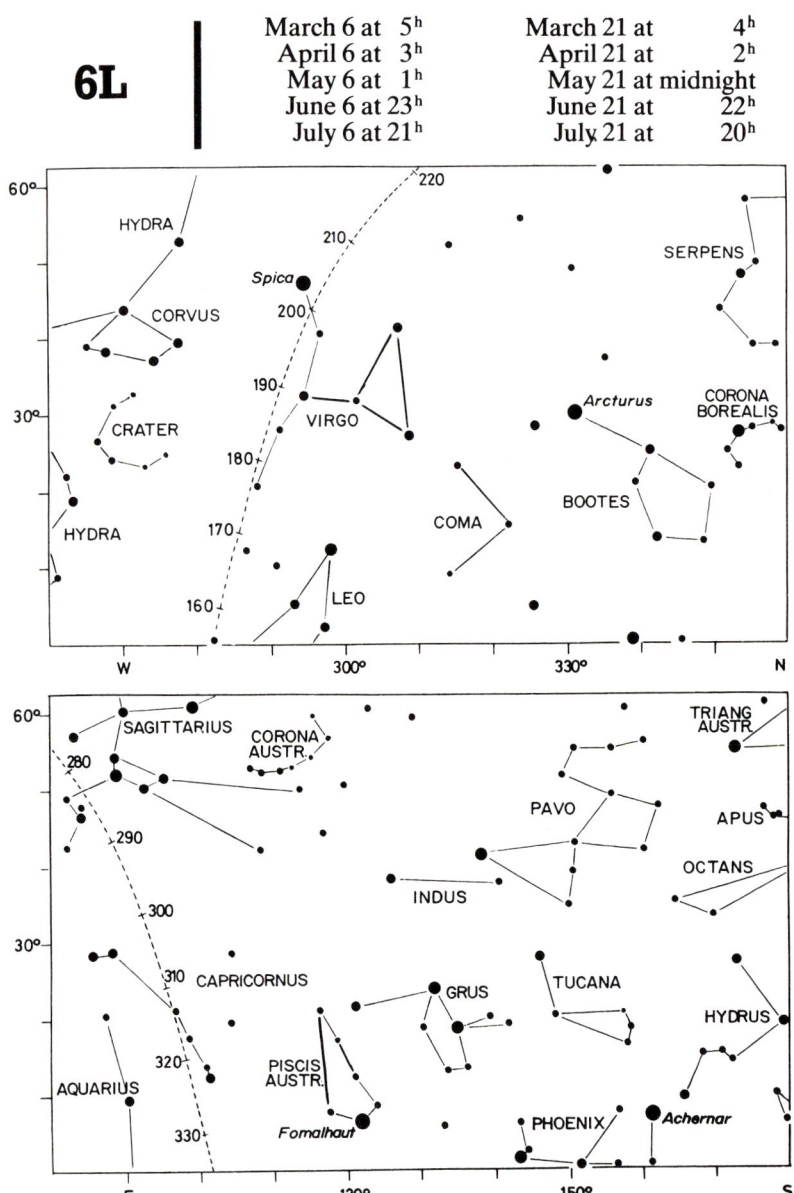

March 6 at 5ʰ	March 21 at 4ʰ	
April 6 at 3ʰ	April 21 at 2ʰ	
May 6 at 1ʰ	May 21 at midnight	**6R**
June 6 at 23ʰ	June 21 at 22ʰ	
July 6 at 21ʰ	July 21 at 20ʰ	

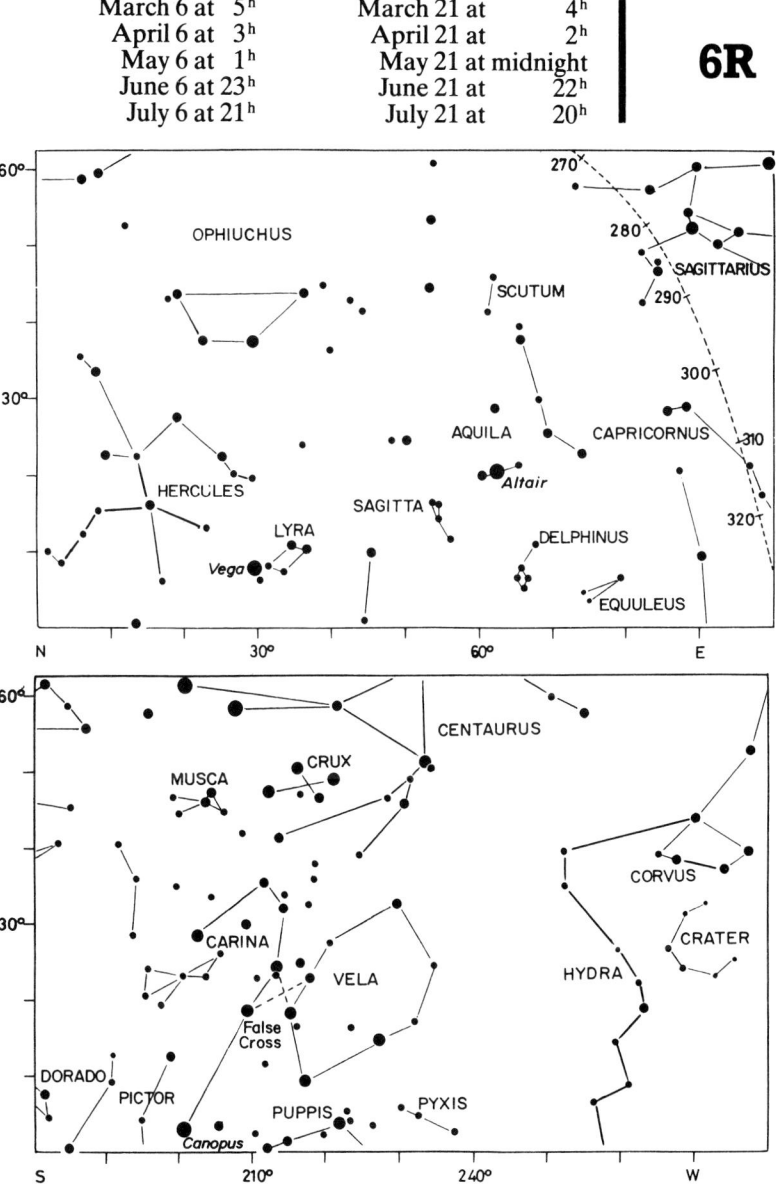

7L

April 6 at 5ʰ   April 21 at   4ʰ
May 6 at 3ʰ   May 21 at   2ʰ
June 6 at 1ʰ   June 21 at midnight
July 6 at 23ʰ   July 21 at   22ʰ
August 6 at 21ʰ   August 21 at   20ʰ

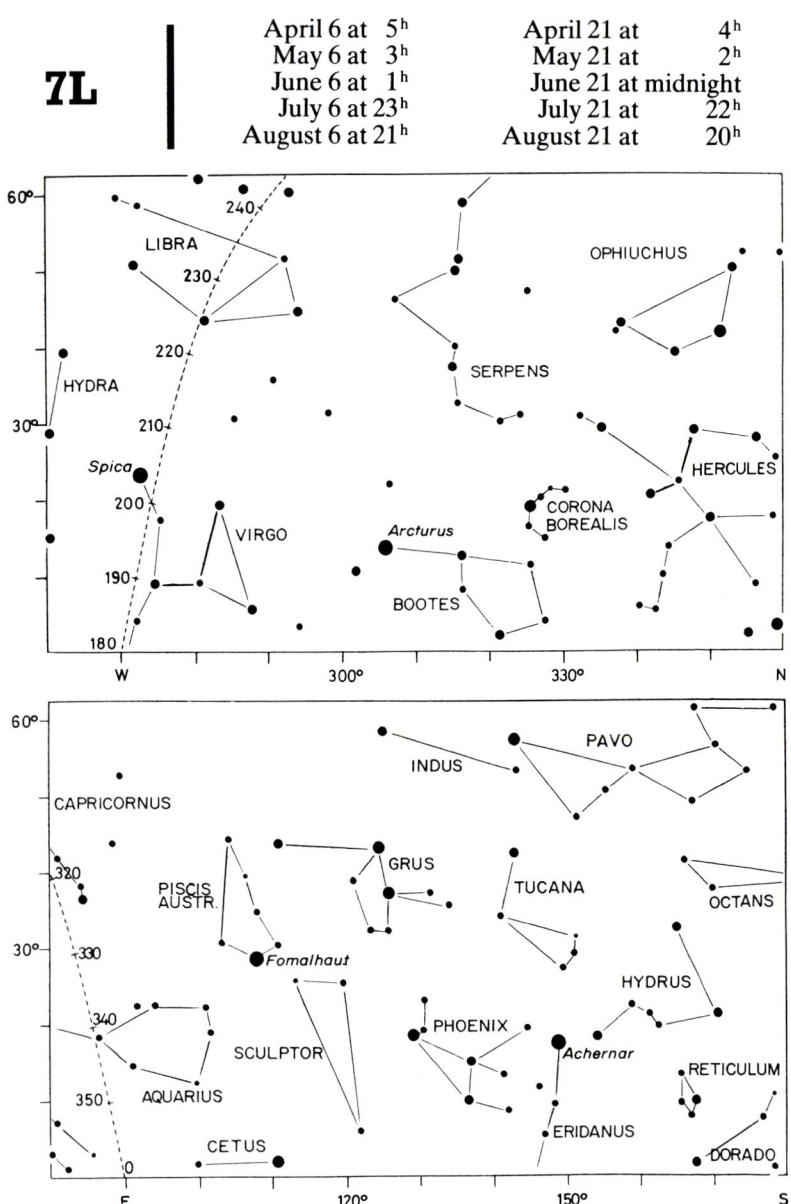

April 6 at 5ʰ   April 21 at   4ʰ
May 6 at 3ʰ   May 21 at   2ʰ
June 6 at 1ʰ   June 21 at midnight
July 6 at 23ʰ   July 21 at   22ʰ
August 6 at 21ʰ   August 21 at   20ʰ

7R

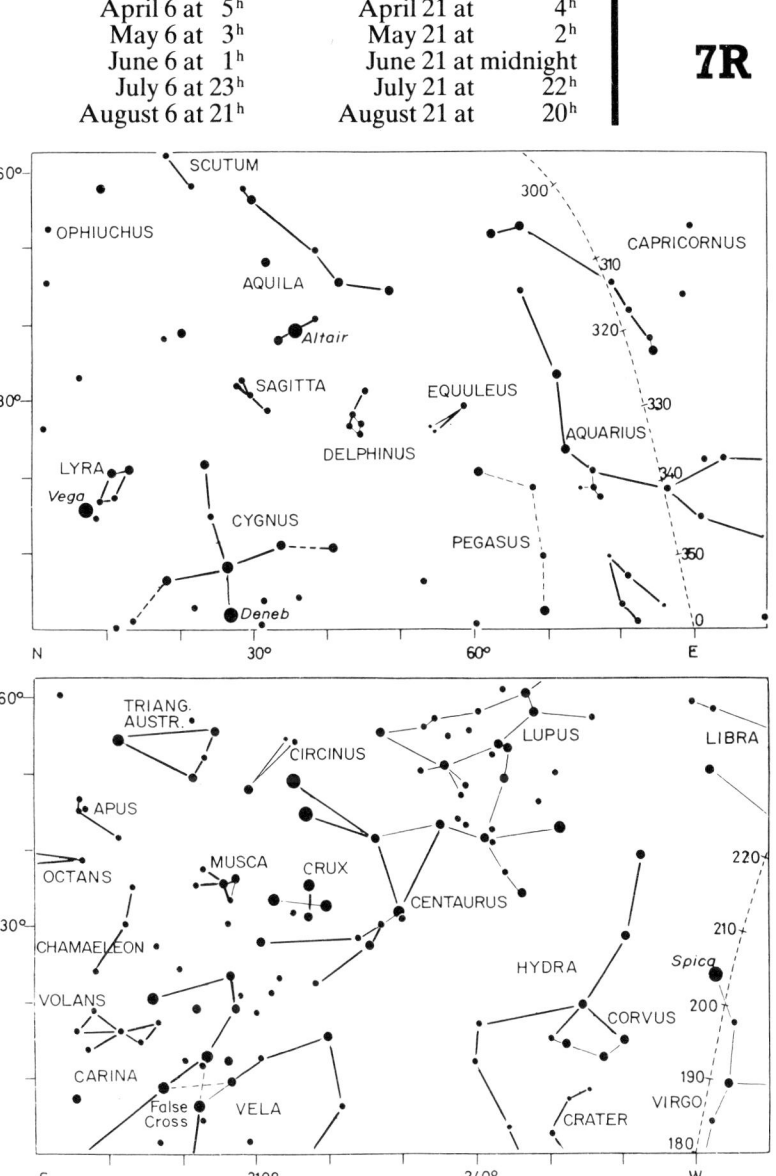

8L

May 6 at 5^h	May 21 at 4^h
June 6 at 3^h	June 21 at 2^h
July 6 at 1^h	July 21 at midnight
August 6 at 23^h	August 21 at 22^h
September 6 at 21^h	September 21 at 20^h

May 6 at 5ʰ
June 6 at 3ʰ
July 6 at 1ʰ
August 6 at 23ʰ
September 6 at 21ʰ

May 21 at 4ʰ
June 21 at 2ʰ
July 21 at midnight
August 21 at 22ʰ
September 21 at 20ʰ

8R

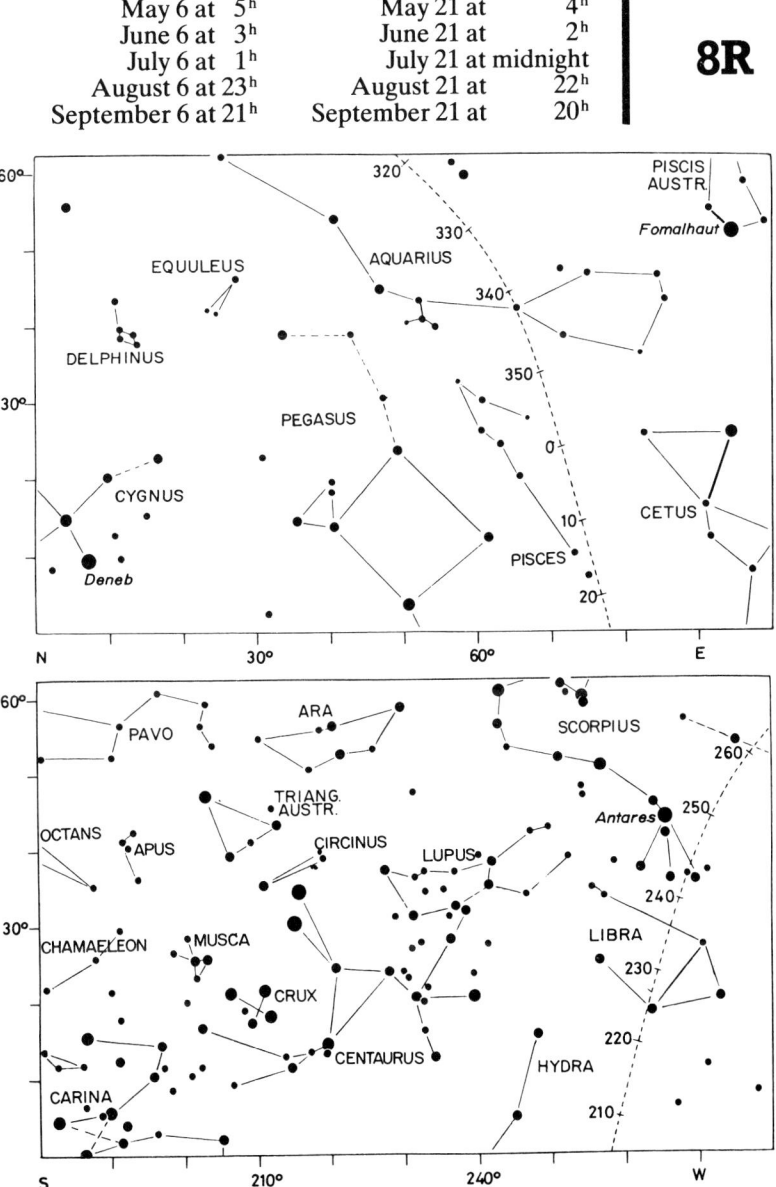

9L

June 6 at 5^h	June 21 at 4^h
July 6 at 3^h	July 21 at 2^h
August 6 at 1^h	August 21 at midnight
September 6 at 23^h	September 21 at 22^h
October 6 at 21^h	October 21 at 20^h

June 6 at	5^h	June 21 at	4^h	
July 6 at	3^h	July 21 at	2^h	**9R**
August 6 at	1^h	August 21 at midnight		
September 6 at	23^h	September 21 at	22^h	
October 6 at	21^h	October 21 at	20^h	

10L

July 6 at 5ʰ	July 21 at 4ʰ
August 6 at 3ʰ	August 21 at 2ʰ
September 6 at 1ʰ	September 21 at midnight
October 6 at 23ʰ	October 21 at 22ʰ
November 6 at 21ʰ	November 21 at 20ʰ

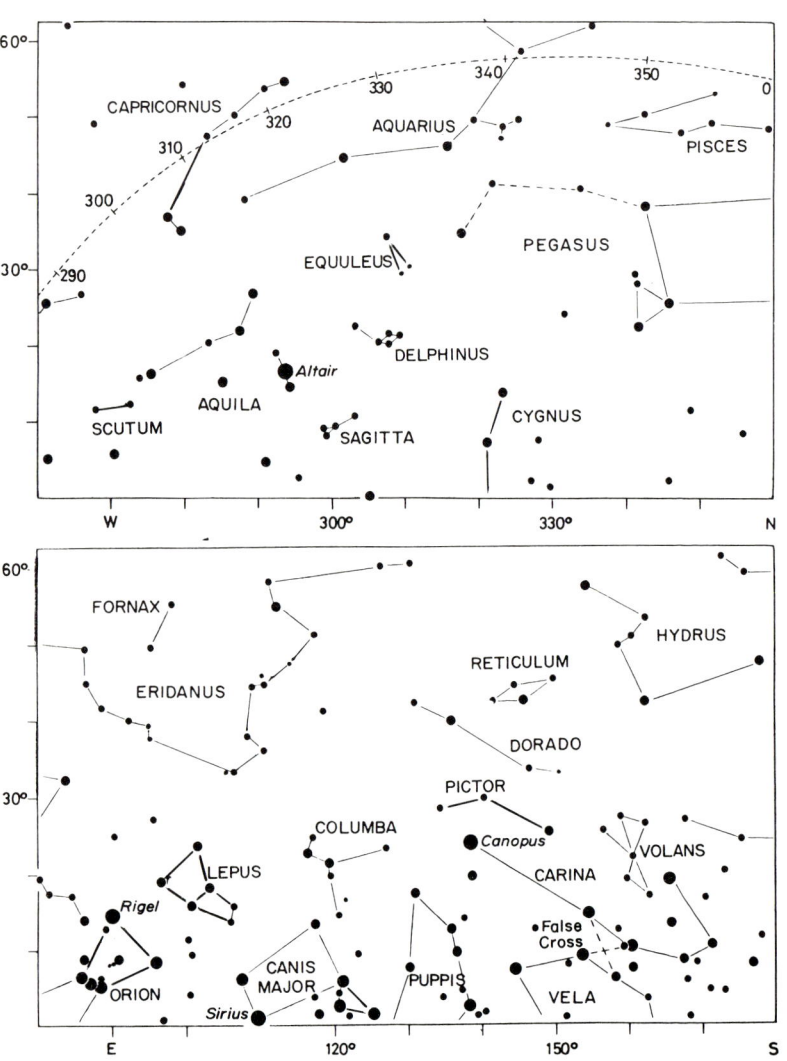

July 6 at 5ʰ	July 21 at 4ʰ	
August 6 at 3ʰ	August 21 at 2ʰ	**10R**
September 6 at 1ʰ	September 21 at midnight	
October 6 at 23ʰ	October 21 at 22ʰ	
November 6 at 21ʰ	November 21 at 20ʰ	

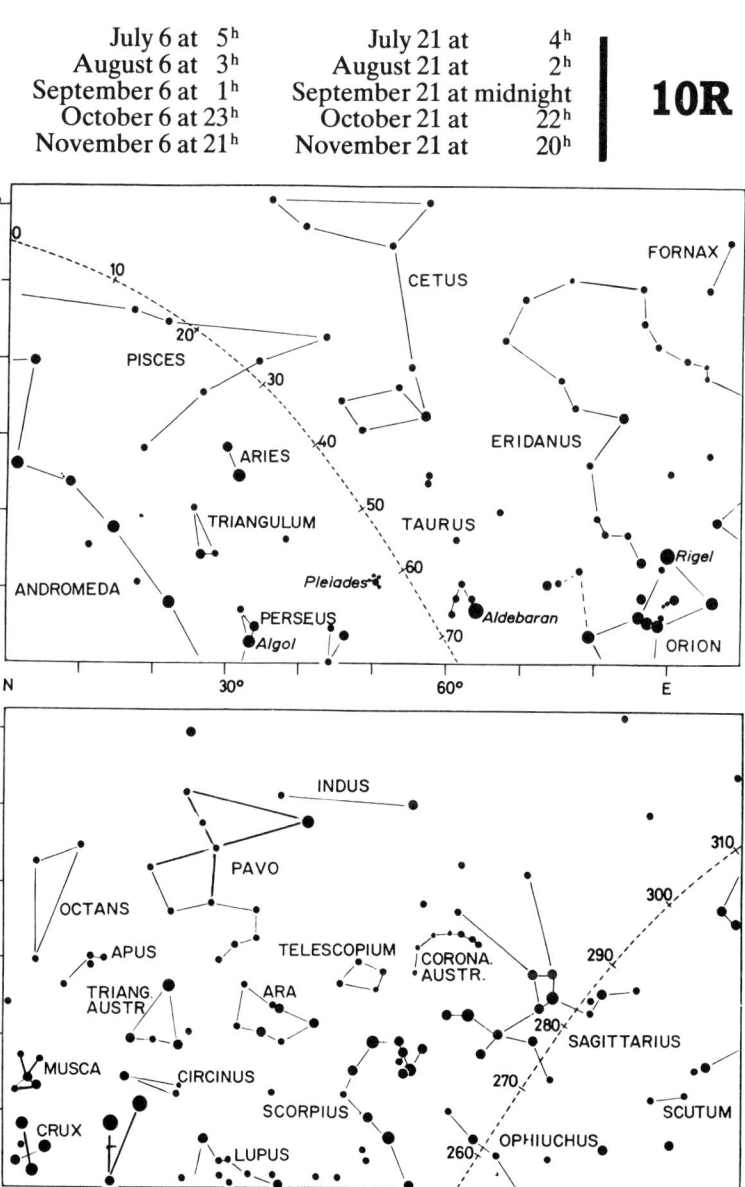

11L

August 6 at 5ʰ	August 21 at 4ʰ
September 6 at 3ʰ	September 21 at 2ʰ
October 6 at 1ʰ	October 21 at midnight
November 6 at 23ʰ	November 21 at 22ʰ
December 6 at 21ʰ	December 21 at 20ʰ

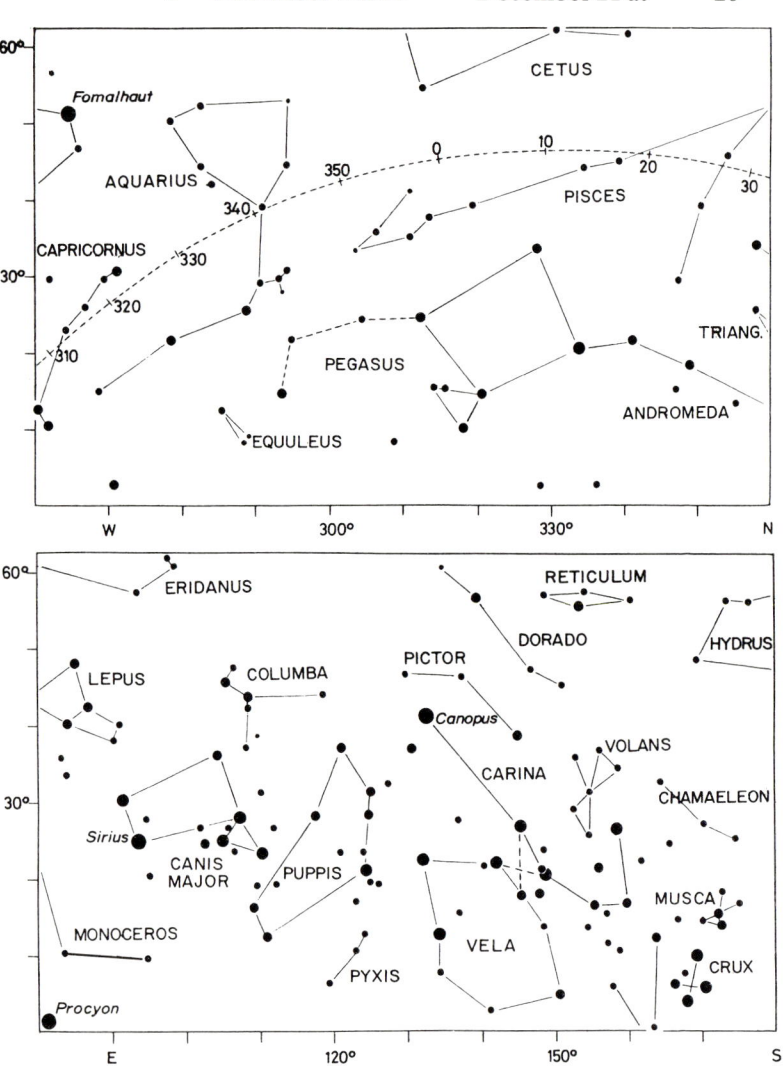

August 6 at 5^h	August 21 at 4^h	
September 6 at 3^h	September 21 at 2^h	**11R**
October 6 at 1^h	October 21 at midnight	
November 6 at 23^h	November 21 at 22^h	
December 6 at 21^h	December 21 at 20^h	

12L

September 6 at 5ʰ | September 21 at 4ʰ
October 6 at 3ʰ | October 21 at 2ʰ
November 6 at 1ʰ | November 21 at midnight
December 6 at 23ʰ | December 21 at 22ʰ
January 6 at 21ʰ | January 21 at 20ʰ

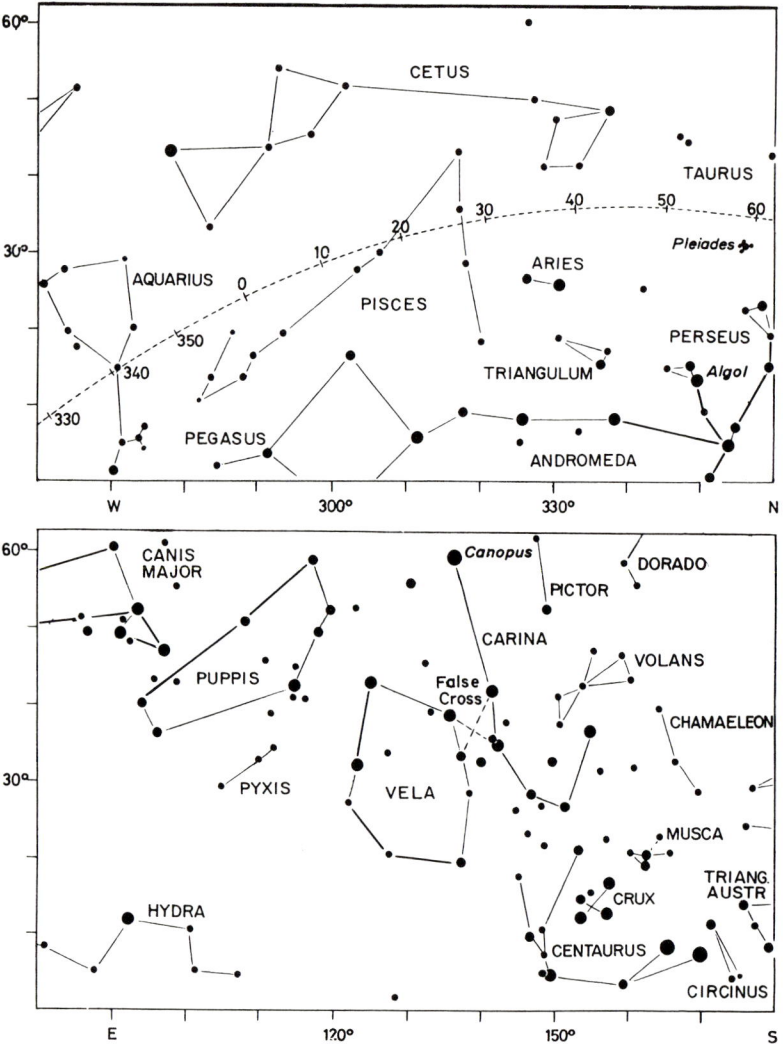

September 6 at 5^h	September 21 at 4^h	
October 6 at 3^h	October 21 at 2^h	**12R**
November 6 at 1^h	November 21 at midnight	
December 6 at 23^h	December 21 at 22^h	
January 6 at 21^h	January 21 at 20^h	

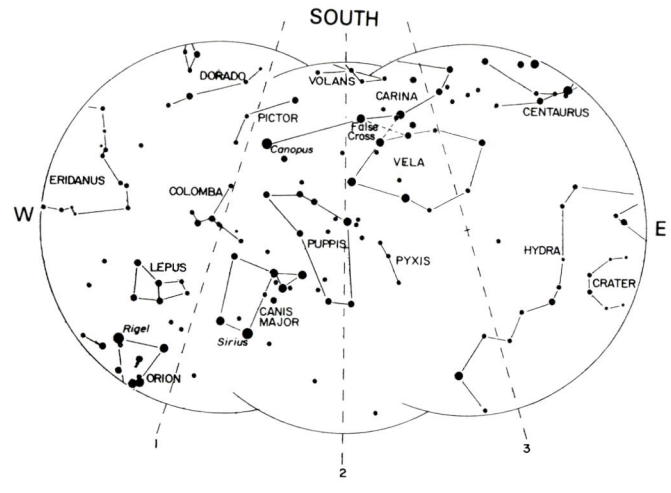

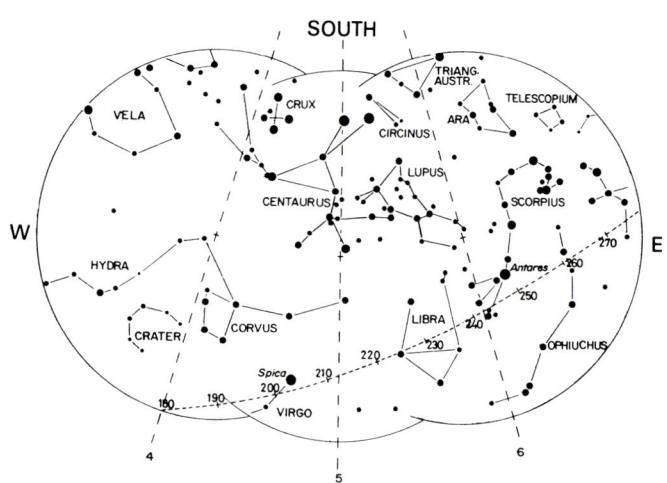

Southern Hemisphere Overhead Stars

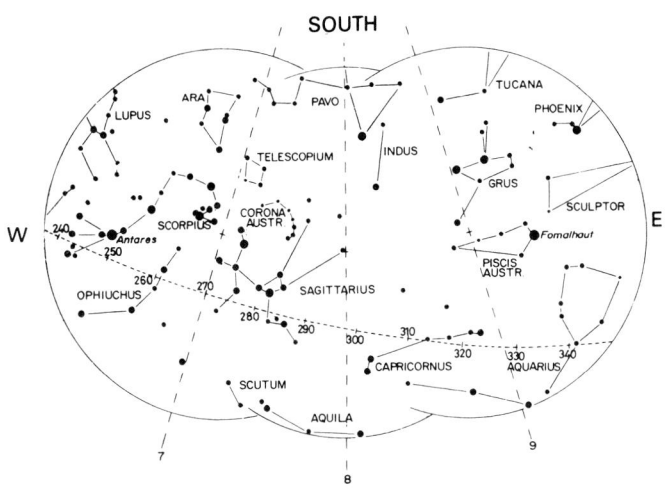

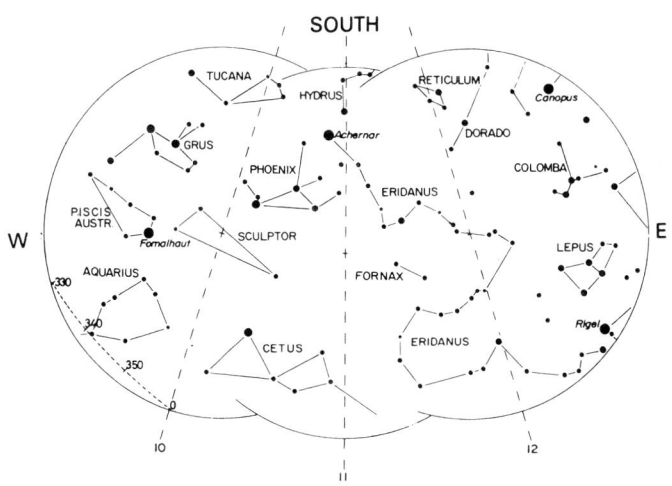

Southern Hemisphere Overhead Stars

The Planets and the Ecliptic

The paths of the planets about the Sun all lie close to the plane of the ecliptic, which is marked for us in the sky by the apparent path of the Sun among the stars, and is shown on the star charts by a broken line. The Moon and planets will always be found close to this line, never departing from it by more than about 7 degrees. Thus the planets are most favourably placed for observation when the ecliptic is well displayed, and this means that it should be as high in the sky as possible. This avoids the difficulty of finding a clear horizon, and also overcomes the problem of atmospheric absorption, which greatly reduces the light of the stars. Thus a star at an altitude of 10 degrees suffers a loss of 60 per cent of its light, which corresponds to a whole magnitude; at an altitude of only 4 degrees, the loss may amount to two magnitudes.

The position of the ecliptic in the sky is therefore of great importance, and since it is tilted at about 23½ degrees to the Equator, it is only at certain times of the day or year that it is displayed to the best advantage. It will be realized that the Sun (and therefore the ecliptic) is at its highest in the sky at noon in midsummer, and at its lowest at noon in midwinter. Allowing for the daily motion of the sky, these times lead to the fact that the ecliptic is highest at midnight in winter, at sunset in the spring, at noon in summer and at sunrise in the autumn. Hence these are the best times to see the planets. Thus, if Venus is an evening object, in the western sky after sunset, it will be seen to best advantage if this occurs in the spring, when the ecliptic is high in the sky and slopes down steeply to the horizon. This means that the planet is not only higher in the sky, but will remain for a much longer period above the horizon. For similar reasons, a morning object will be seen at its best on autumn mornings before sunrise, when the ecliptic is high in the east. The outer planets, which can come to opposition (i.e. opposite the Sun), are best seen when opposition occurs in the winter months, when the ecliptic is high in the sky at midnight.

The seasons are reversed in the Southern Hemisphere, spring beginning at the September Equinox, when the Sun crosses the Equator on its way south, summer beginning at the December

Solstice, when the Sun is highest in the southern sky, and so on. Thus, the times when the ecliptic is highest in the sky, and therefore best placed for observing the planets, may be summarized as follows:

	Midnight	*Sunrise*	*Noon*	*Sunset*
Northern lats.	December	September	June	March
Southern lats.	June	March	December	September

In addition to the daily rotation of the celestial sphere from east to west, the planets have a motion of their own among the stars. The apparent movement is generally *direct*, i.e. to the east, in the direction of increasing longitude, but for a certain period (which depends on the distance of the planet) this apparent motion is reversed. With the outer planets this *retrograde* motion occurs about the time of opposition. Owing to the different inclination of the orbits of these planets, the actual effect is to cause the apparent path to form a loop, or sometimes an S-shaped curve. The same effect is present in the motion of the inferior planets, Mercury and Venus, but it is not so obvious, since it always occurs at the time of inferior conjunction.

The inferior planets, Mercury and Venus, move in smaller orbits than that of the Earth, and so are always seen near the Sun. They are most obvious at the times of greatest angular distance from the Sun (greatest elongation), which may reach 28 degrees for Mercury, or 47 degrees for Venus. They are seen as evening objects in the western sky after sunset (at eastern elongations) or as morning objects in the eastern sky before sunrise (at western elongations). The succession of phenomena, conjunctions and elongations, always follows the same order, but the intervals between them are not equal. Thus, if either planet is moving round the far side of its orbit its motion will be to the east, in the same direction in which the Sun appears to be moving. It therefore takes much longer for the planet to overtake the Sun – that is, to come to superior conjunction – than it does when moving round to inferior conjunction, between Sun and Earth. The intervals given in the following table are average values; they remain fairly constant in the case of Venus, which travels in an almost circular orbit. In the case of Mercury, however, conditions vary widely because of the great eccentricity and inclination of the planet's orbit.

		Mercury	*Venus*
Inferior conj.	to Elongation West	22 days	72 days
Elongation West	to Superior conj.	36 days	220 days
Superior conj.	to Elongation East	36 days	220 days
Elongation East	to Inferior conj.	22 days	72 days

The greatest brilliancy of Venus always occurs about 36 days before or after inferior conjunction. This will be about a month *after* greatest eastern elongation (as an evening object), or a month *before* greatest western elongation (as a morning object). No such rule can be given for Mercury, because its distance from the Earth and the Sun can vary over a wide range.

Mercury is not likely to be seen unless a clear horizon is available. It is seldom seen as much as 10 degrees above the horizon in the twilight sky in northern latitudes, but this figure is often exceeded in the Southern Hemisphere. This favourable condition arises because the maximum elongation of 28 degrees can occur only when the planet is at aphelion (farthest from the Sun), and this point lies well south of the Equator. Northern observers must be content with smaller elongations, which may be as little as 18 degrees at perihelion. In general, it may be said that the most favourable times for seeing Mercury as an evening object will be in spring, some days before greatest eastern elongation; in autumn, it may be seen as a morning object some days after greatest western elongation.

Venus is the brightest of the planets and may be seen on occasions in broad daylight. Like Mercury, it is alternately a morning and an evening object, and it will be highest in the sky when it is a morning object in autumn, or an evening object in spring. The phenomena of Venus given in the table above can occur only in the months of January, April, June, August and November, and it will be realized that they do not all lead to favourable apparitions of the planet. In fact, Venus is to be seen at its best as an evening object in northern latitudes when eastern elongation occurs in June. The planet is then well north of the Sun in the preceding spring months, and is a brilliant object in the evening sky over a long period. In the Southern Hemisphere a November elongation is best. For similar reasons, Venus gives a prolonged display as a morning object in the months following western elongation in November (in northern latitudes) or in June (in the Southern Hemisphere).

The superior planets, which travel in orbits larger than that of the Earth, differ from Mercury and Venus in that they can be seen opposite the Sun in the sky. The superior planets are morning objects after conjunction with the Sun, rising earlier each day until they come to opposition. They will then be nearest to the Earth (and therefore at their brightest), and will then be on the meridian at midnight, due south in northern latitudes, but due north in the Southern Hemisphere. After opposition they are evening objects, setting earlier each evening until they set in the west with the Sun at the next conjunction. The change in brightness about the time of opposition is most noticeable in the case of Mars, whose distance from Earth can vary considerably and rapidly. The other superior planets are at such great distances that there is very little change in brightness from one opposition to another. The effect of altitude is, however, of some importance, for at a December opposition in northern latitudes the planets will be among the stars of Taurus or Gemini, and can then be at an altitude of more than 60 degrees in southern England. At a summer opposition, when the planet is in Sagittarius, it may only rise to about 15 degrees above the southern horizon, and so makes a less impressive appearance. In the Southern Hemisphere, the reverse conditions apply; a June opposition being the best, with the planet in Sagittarius at an altitude which can reach 80 degrees above the northern horizon for observers in South Africa.

Mars, whose orbit is appreciably eccentric, comes nearest to the Earth at an opposition at the end of August. It may then be brighter even than Jupiter, but rather low in the sky in Aquarius for northern observers, though very well placed for those in southern latitudes. These favourable oppositions occur every fifteen or seventeen years (1956, 1971, 1988, 2003) but in the Northern Hemisphere the planet is probably better seen at an opposition in the autumn or winter months, when it is higher in the sky. Oppositions of Mars occur at an average interval of 780 days, and during this time the planet makes a complete circuit of the sky.

Jupiter is always a bright planet, and comes to opposition a month later each year, having moved, roughly speaking, from one Zodiacal constellation to the next.

Saturn moves much more slowly than Jupiter, and may remain in the same constellation for several years. The brightness of Saturn depends on the aspects of its rings, as well as on the distance from Earth and Sun. The rings were inclined towards the Earth and Sun in 1980 and are currently near their maximum opening. The next passage of both Earth and Sun through the ring-plane will not occur until 1995.

Uranus, *Neptune*, and *Pluto* are hardly likely to attract the attention of observers without adequate instruments.

Phases of the Moon
1990

New Moon				*First Quarter*				*Full Moon*				*Last Quarter*		
	d	h	m		d	h	m		d	h	m		d	h	m
Jan.	26	19	20	Jan.	4	10	40	Jan.	11	04	57	Jan.	18	21	17
Feb.	25	08	54	Feb.	2	18	32	Feb.	9	19	16	Feb.	17	18	48
Mar.	26	19	48	Mar.	4	02	05	Mar.	11	10	58	Mar.	19	14	30
Apr.	25	04	27	Apr.	2	10	24	Apr.	10	03	18	Apr.	18	07	02
May	24	11	47	May	1	20	18	May	9	19	31	May	17	19	45
June	22	18	55	May	31	08	11	June	8	11	01	June	16	04	48
July	22	02	54	June	29	22	07	July	8	01	23	July	15	11	04
Aug.	20	12	39	July	29	14	01	Aug.	6	14	19	Aug.	13	15	54
Sept.	19	00	46	Aug.	28	07	34	Sept.	5	01	46	Sept.	11	20	53
Oct.	18	15	37	Sept.	27	02	06	Oct.	4	12	02	Oct.	11	03	31
Nov.	17	09	05	Oct.	26	20	26	Nov.	2	21	48	Nov.	9	13	02
Dec.	17	04	22	Nov.	25	13	11	Dec.	2	07	50	Dec.	9	02	04
				Dec.	25	03	16	Dec.	31	18	35				

All times are G.M.T.
Reproduced, with permission, from data supplied by the Science and Engineering
Research Council.

Longitudes of the Sun, Moon and Planets in 1990

DATE		Sun °	Moon °	Venus °	Mars °	Jupiter °	Saturn °
January	6	285	36	305	253	95	286
	21	301	234	298	263	93	288
February	6	317	89	291	275	91	290
	21	332	278	294	286	91	291
March	6	345	100	302	296	91	293
	21	0	286	314	307	92	294
April	6	16	149	330	319	93	295
	21	31	333	345	330	95	295
May	6	45	183	2	341	98	295
	21	60	10	19	352	101	295
June	6	75	228	38	4	104	295
	21	89	64	55	15	107	294
July	6	104	261	73	25	110	293
	21	118	103	90	35	114	292
August	6	133	306	110	46	117	290
	21	148	154	128	54	120	290
September	6	163	355	148	63	124	289
	21	178	201	166	69	127	289
October	6	192	33	186	73	129	289
	21	207	233	204	75	131	289
November	6	223	87	224	73	133	290
	21	238	278	243	68	133	291
December	6	254	125	262	62	134	293
	21	269	311	280	59	133	294

Longitude of *Uranus* 278°
        Neptune 284°

Moon: Longitude of ascending node
        Jan. 1: 318°   Dec. 31: 299°

Mercury moves so quickly among the stars that it is not possible to indicate its position on the star charts at a convenient interval. The monthly notes must be consulted for the best times at which the planet may be seen.

The positions of the other planets are given in the table on the previous page. This gives the apparent longitudes on dates which correspond to those of the star charts, and the position of the planet may at once be found near the ecliptic at the given longitude.

Examples
From the southern hemisphere two planets are seen in the eastern morning sky in March. Identify them.

> The southern star chart 5L shows the eastern sky at March 6^d3^h and shows longitudes $255° - 310°$. Reference to the table on page 71 gives the longitude of Mars as 296° and that of Saturn as 293°, on February 6. Thus these planets are found below the main stars of Sagittarius and the one with the slightly reddish tint is Mars.

The positions of the Sun and Moon can be plotted on the star maps in the same manner as for the planets. The average daily motion of the Sun is 1°, and of the Moon 13°. For the Moon an indication of its position relative to the ecliptic may be obtained from a consideration of its longitude relative to that of the ascending node. The latter changes only slowly during the year as will be seen from the values given on the previous page. Let us call the difference in longitude of Moon-node, d. Then if d = 0°, 180° or 360° the Moon is on the ecliptic. If d = 90° the Moon is 5° north of the ecliptic and if d = 270° the Moon is 5° south of the ecliptic.

On October 6 the Moon's longitude is given as 33° and the longitude of the node is found by interpolation to be about 303°. Thus d = 90° and the Moon is about 5° north of the ecliptic. Its position may be plotted on northern star charts 1R, 2L, 2R, 8R, 9L, 10L and 12R: and southern star charts 1L, 9R, 10R and 12L.

Events in 1990

ECLIPSES (See page 110)
In 1990 there will be four eclipses, two of the Sun and two of the
Moon.

January 26: annular eclipse of the Sun – the partial phase being
visible from south of New Zealand, Antarctica, S. America (except
N.W.).

February 9: total eclipse of the Moon – N.W. Alaska, arctic
regions, Australasia, Asia, Africa, Europe, Iceland, Greenland.

July 22: total eclipse of the Sun – the partial phase being visible
from N.E. Europe, N. of Greenland, N. Asia, Arctic regions, N.W.
of N. America, Hawaiian Islands.

August 6: partial eclipse of the Moon – S.W. Alaska, Pacific
Ocean, Antarctica, Australasia, S. and E. Asia.

THE PLANETS
Mercury may be seen more easily from northern latitudes in the
evenings about the time of greatest eastern elongation (April 13)
and in the mornings around greatest western elongation (September 24). In the southern hemisphere the corresponding dates are
February 1 (mornings) and August 11 (evenings). Venus is visible in
the evenings for the first part of January. From late in January until
late in September it is visible in the mornings. It is again visible in
the evenings for the second half of December.

Mars is a morning object until opposition on November 27: thereafter it is an evening object.

Jupiter is not at opposition until 1991 January 29

Saturn is at opposition on July 14

Uranus is at opposition on June 29

Neptune is at opposition on July 5

Pluto is at opposition on May 7

JANUARY

Full Moon: January 11   *New Moon:* January 26

EARTH is at perihelion (nearest to the Sun) on January 4 at a distance of 147 million kilometres.

MERCURY is in inferior conjunction on January 9 and comes to greatest western elongation on February 1. Observers in equatorial and southern latitudes may be able to see the planet as a difficult evening object for the first two or three days of the month. For these observers Mercury will have re-appeared as a morning object by the middle of January. Those in northern temperate latitudes will only have any hope of seeing Mercury during the last ten days of the month, low above the south-eastern horizon before dawn.

VENUS, for the first half of the month, is visible in the evenings as a brilliant object low above the west to south-western horizon for a short while after sunset. Venus passes through inferior conjunction on January 18 and therefore will not be visible for some days either side of that date. However, Venus is moving rapidly with respect to the Sun and is clearly visible in the mornings for the last week of the month, low above the east-south-eastern horizon for a short while before sunrise. A keen observer might note that Venus is not actually on the ecliptic. In fact by the end of January Venus is as much as $7°.3$ N of it. This is due to its proximity to the Earth: its *heliocentric* latitude is only $+2°.9$.

MARS begins the year in Ophiuchus moving eastwards into Sagittarius before the end of January. It is not a very conspicuous object (magnitude $+1.5$), visible in the south-eastern sky before morning inhibits observation. Its position among the stars during the early months of the year is shown in Figure 3, given with the monthly notes for March.

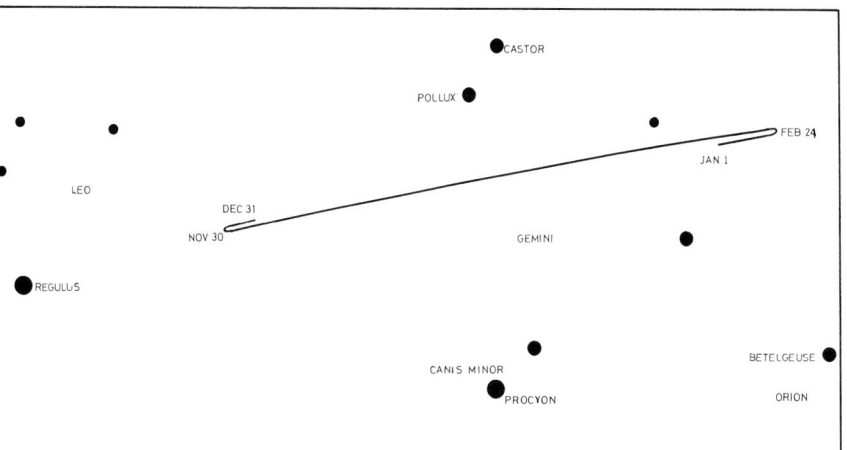

Figure 1. The path of Jupiter.

JUPITER is now just past opposition (December 27) so that it is visible for the greater part of the night. Jupiter is an outstanding object in Gemini, with a magnitude of −2.7. Figure 1 shows the path of Jupiter throughout the year, relative to the stars.

SATURN is in conjunction with the Sun on January 6 and therefore not suitably placed for observation until nearly the end of the month when keen-sighted Southern-Hemisphere observers may be able to glimpse it for a short while in the early mornings low above the south-eastern horizon.

THIS MONTH'S ANNULAR ECLIPSE. If the Earth, Moon and Sun are aligned at a time when the Moon is in the further part of its orbit, the apparent diameter of the Moon is less than that of the Sun, so that a ring of sunlight is left showing round the dark disk of the Moon. This is what far-southern observers will see on January 26. The name annular comes from the Latin *annulus*, a ring.

Annular eclipses are interesting to observe, but they are not spectacular, and those people who have not been told may well overlook such an eclipse altogether; the reduction in sunlight is marked, but not noticeably so. Moreover, the phenomena of totality – the chromosphere, corona and prominences – cannot be seen, though they have been glimpsed when the Moon is almost big enough to hide the Sun completely (as will not be the case this month). It may be wise to repeat the usual **warning: never look at the Sun direct with any optical appliance, as damage to the observer's eyesight is certain to occur**.

Total eclipses are always hurried affairs. Annular eclipses may be more leisurely, and may last for as long as 12 minutes 24 seconds. For photographic hints, see the article by Michael Maunder in this *Yearbook*.

Incidentally, the Earth is at perihelion this month, so that the Sun's apparent diameter is at its greatest: 32 minutes 35 seconds of arc. At aphelion, the apparent diameter is only 31 minutes 31

An annular eclipse

seconds. For comparison, the apparent diameter of the Moon ranges between 33 minutes 31 seconds (perigee) and 29 minutes 22 seconds (apogee). There is no other case in the Solar System where the agreement between the apparent diameter of the Sun and a satellite is so good, though there are many satellites which can cause total solar eclipses as seen from their primaries: Amalthea and all four Galilean satellites from Jupiter; Mimas, Enceladus, Tethys, Rhea, Dione and Titan from Saturn; Miranda, Ariel, Umbriel, Titania and Oberon from Uranus; and Triton from Neptune.

THIS MONTH'S CENTENARY. On January 23, 1890 occurred the death of the Latvian astronomer Otto August Rosenberger. He was born at Tukkim, Latvia, on August 10, 1800; he became assistant to Bessel at Königsberg, and later Professor at Halle. He is best remembered, perhaps, for his very accurate predictions of the perihelion of Halley's Comet at the return of 1835.

FEBRUARY

MERCURY, although it reaches greatest western elongation (25°) on the first day of the month, is not suitably placed for observation by those in northern temperate latitudes. For observers further south this is the most favourable morning apparition of the year as Mercury may be seen from the middle of January until early in March. Figure 2 shows, for observers in latitude S.35°, the changes in azimuth (true bearing from north through east, south and west) and altitude of Mercury on successive mornings when the Sun is 6° below the horizon. This condition is known as the beginning of morning civil twilight, and in this latitude and at this time of year occurs about 30 minutes before sunrise. The changes in the brightness of the planet are indicated by the relative sizes of the circles marking Mercury's positions at five-day intervals. It will be noticed that Mercury is brightest after it reaches greatest western elongation. During the first half of the month inexperienced observers may be misled by the appearance of Venus in the same part of the sky. Venus is 5 magnitudes brighter than Mercury.

VENUS is a brilliant object in the mornings, completely dominating the eastern skies before dawn. Venus attains its greatest brilliancy on February 22 (magnitude −4.6). It is a beautiful sight in a small telescope at this time, as it exhibits a slender crescent phase. During the month the phase increases noticeably as the apparent diameter decreases.

MARS is a morning object, magnitude +1.3, in the south-eastern sky before dawn. Observers in the latitudes of the British Isles will continue to have difficulty in locating it because of its low altitude.

JUPITER is an evening object, though still visible for several hours after midnight.

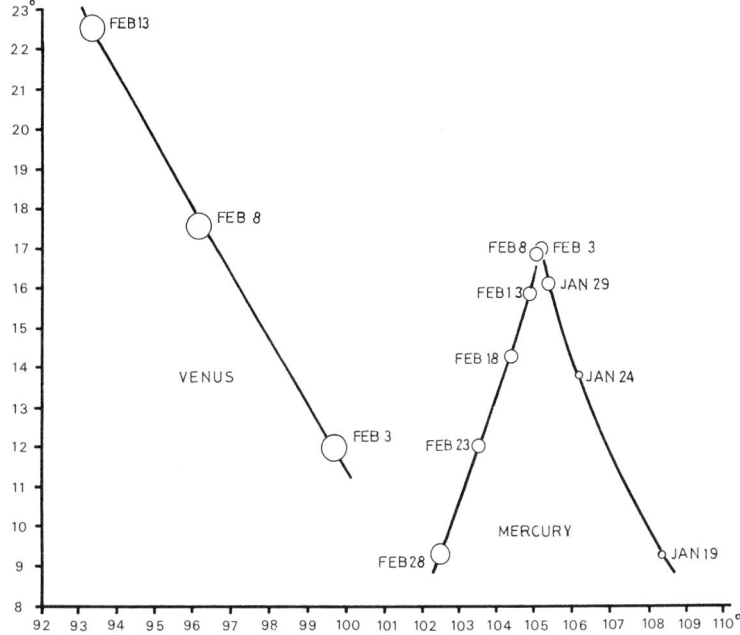

Figure 2. Morning apparition of Mercury for latitude S.35°.

SATURN is a difficult morning object, magnitude +0.6, visible low in the south-eastern sky for a short while before twilight inhibits observation – though this is not true for observers in northern temperate latitudes who will have to wait until March before they can hope to see the planet again.

ETA CARINÆ. During February evenings the constellation of Carina, the Keel, is high in the sky as seen from South Africa, Australia or New Zealand; it is dominated by the brilliant Canopus, which is the brightest star in the sky with the exception of Sirius, and is a true 'cosmic searchlight'. Estimates of its luminosity vary, but according to the authoritative Cambridge catalogue it must be the equal of 200,000 Suns. Its spectral type is F, so that it should appear slightly yellowish, though most people see it as white.

Yet even Canopus is outmatched by another star in Carina. This is Eta Carinæ, the most erratic variable in the sky, and unlike anything else known to us. It was described by D. A. Allen in a past *Yearbook*, but very recently some new information about it has been obtained.

Today, Eta Carinæ is just below naked-eye visibility; any binoculars will show it, and it appears unusual – it has been likened to 'an orange blob'. For a while, around 1840, it exceeded Canopus in brightness. Earlier it had been easily visible with the naked eye; for instance Edmond Halley, in 1677, recorded it as being of the fourth magnitude. After its brilliant period, a century and a half ago, it declined, though it has never fallen below the seventh magnitude.

Eta Carinæ is very distant, and immensely luminous. It must be at least 5,000,000 times as powerful as the Sun, and perhaps as much as 6,000,000 Sun-power, so that it may lay claim to being the most luminous star known to us. It has a spectrum which cannot be put into any definite class, and is given officially as 'Peculiar'. It is a strong infrared source, and is associated with nebulosity; it may be that it is apparently dimmer now than it used to be owing to a greater amount of intervening interstellar matter.

It has been suggested that Eta Carinæ may be a group of stars rather than a single object; this has been found to be the case with, for instance, R136a in the Tarantula Nebula in the Large Cloud of Magellan. Using the 2.2-metre telescope at the La Silla Observatory in Chile, the German astronomers K. H. Hofmann and G. Weigelt have made a close study at near-infrared wavelengths, and have found that there are indeed four components: one dominant star, and three companions at distances of between 0.1 and 0.2 seconds of arc. All three companions are around 12 times fainter than the primary, which still makes the senior member of the group exceptionally luminous.

Another important factor is the exceptionally large mass and the chemical composition of Eta Carinæ. It may be verging on disaster – and it could well be a supernova candidate. The outburst may be delayed for a million years or more, but sooner or later it is likely to happen. When it does, Eta Carinæ will indeed have a brief period of unrivalled glory!

VENUS AT ITS BRIGHTEST. This month Venus is at its maximum brilliancy. It is in its crescent stage, and very keen-eyed observers may care to see whether they can make out the crescent without optical aid. It is very difficult, but there is good evidence that a few people can manage it under near-perfect conditions.

MARCH

Full Moon: March 11   *New Moon:* March 26

Summer Time in Great Britain and Northern Ireland commences on March 25.

Equinox: March 20

MERCURY may be seen by those in equatorial and southern latitudes as a morning object during the first week of the month. See Figure 2 given with the notes for February. Thereafter it is too close to the Sun for observation as superior conjunction occurs on the 19th.

VENUS is still a brilliant object in the morning skies, its magnitude being −4.5. It is south of the Equator and observers in the British Isles will only be able to see it for about an hour before sunrise. Observers in the Southern Hemisphere are more fortunate, enjoying a visibility period of around three hours before dawn. Venus is at greatest western elongation (46°) on March 30.

MARS is continuing its eastward progress, passing into Capricornus during the month. It is a morning object in the south-eastern skies though still not an easy object to locate by observers in the latitudes of the British Isles. Their best chance of seeing the planet is roughly an hour before sunrise. The path of Mars among the stars is shown in Figure 3.

JUPITER continues to be visible as a brilliant evening object, magnitude −2.3, in the constellation of Gemini. For observers in northern temperate latitudes it is still visible well after midnight in the western sky.

SATURN, already visible to observers further south, slowly

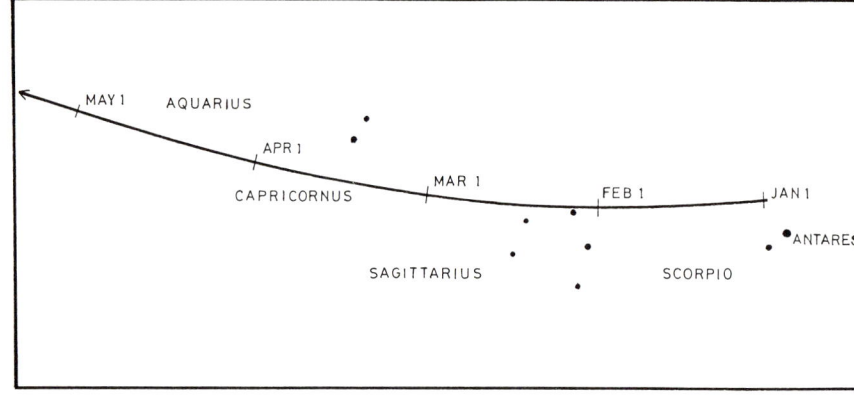

Figure 3. The path of Mars.

becomes visible low in the south-eastern sky in the mornings, to observers in the latitudes of the British Isles. Saturn's magnitude is +0.6. It is in Sagittarius as will be seen in Figure 5, given with the monthly notes for July.

SECULAR VARIABLES. Variable stars of many kinds; some regular, some erratic. There are also a few cases of stars which have been said to brighten or fade permanently over the centuries – the so-called secular variables. Three which have been suspected of fading are well on view this month: Castor in Gemini, Megrez in Ursa Major, and Denebola in Leo.

Yet the evidence does not seem really strong when it is examined closely. Any casual glance at the Great Bear or Plough will show that one of the seven stars – Megrez or Delta Ursæ Majoris – is much fainter than the rest; it is below the third magnitude, and so is about a magnitude below any of the other members of the pattern. Hipparchus (around BC 130) and Ptolemy (around AD 140) agreed with this, making Megrez of magnitude 3 and the rest of magnitude 2. Of the famous star-cataloguers, only Tycho (1590) and Bayer (1603) made Megrez of the 2nd magnitude. Therefore, evidence for any secular variation is decidedly slender.

Castor, one of the Twins, is half a magnitude fainter than Pollux – but up to the time of Flamsteed (1700) both the Twins were given as being of magnitude 2. Only Flamsteed himself ranked Castor as the brighter. Again the evidence is poor; but if there is any real possibility of a change, it is more likely that Pollux has brightened up rather than that Castor has declined. Pollux, remember, is a K-type star, whereas Castor is a multiple system whose main components are of type A.

The case of Denebola, or Beta Leonis, is slightly more convincing, because all observers up to the eighteenth century ranked it as being of the first magnitude, as is Regulus. Flamsteed made Regulus of magnitude 1, Denebola as 1½; since then there has never been any doubt that Denebola is much the fainter of the two, though admittedly it has been suspected of slight fluctuations in modern times. It is 39 light-years away, with an A3 spectrum. Those observers who are interested in naked-eye estimations may care to keep watch on it!

SUMMER TIME begins this month. Time in general is the subject of a separate article in this *Yearbook*, by Gordon Taylor. Remember, however, never to use summer time for recording astronomical observations: always use G.M.T.

APRIL

Full Moon: April 10   *New Moon:* April 25

MERCURY is visible as an evening object until the last few days of the month. For observers in northern temperate latitudes this will be the most favourable evening apparition of the year, although for them the visibility period is restricted to the two middle weeks of the month. Figure 4 shows, for observers in latitudes N.52°, the changes in azimuth (true bearing from the north through east, south and west) and altitude of Mercury on successive evenings when the Sun is 6° below the horizon. This condition is known as the end of evening civil twilight and in this latitude and at this time of year occur about 40 minutes after sunset. The changes in the brightness of the planet are indicated by the relative sizes of the circles marking Mercury's positions at five-day intervals. It will be noticed that Mercury is brightest before it reaches greatest eastern elongation on April 13. Its magnitude on April 9 is −0.5 while by April 19 it is only +1.1.

VENUS continues to be visible as a brilliant object in the mornings before dawn. However, it is only visible to observers in northern temperate latitudes for a short while before dawn, above the east-south-eastern horizon. Southern Hemisphere observers will continue to enjoy a three-hour period of visibility.

MARS is still a morning object, in the south-eastern sky, magnitude +0.9.

JUPITER is still a brilliant evening object, magnitude −2.1.

SATURN is a morning object, magnitude +0.6, in the constellation of Sagittarius.

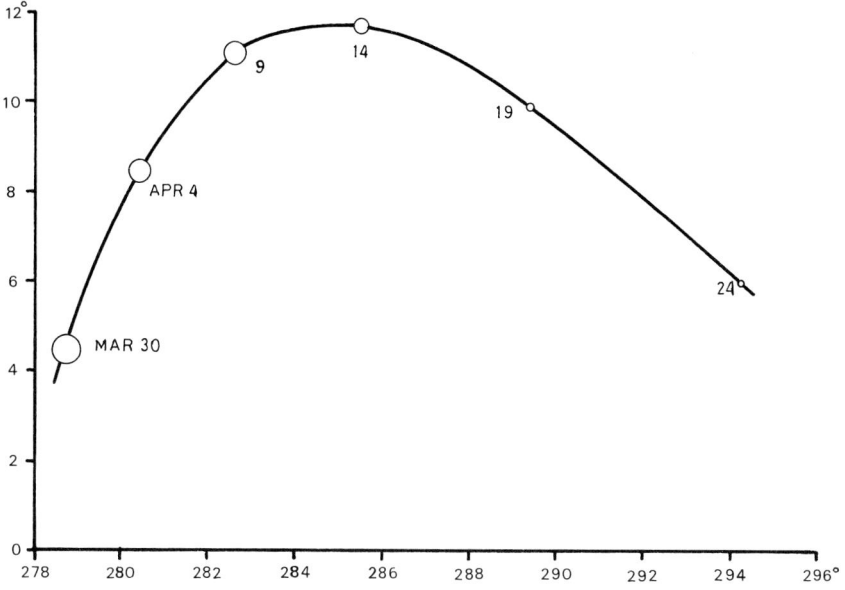

Figure 4. Evening apparition of Mercury for latitude N.52°.

MERCURY, the innermost planet, is never a prominent naked-eye object, because it remains close to the Sun in the sky and can never be seen against a dark background. However, it is actually a bright object, and at its best it can almost equal Sirius in magnitude. For northern observers, this April presents the best chance this year of seeing it.

Of course, telescope users who have suitable equipment can find Mercury in daylight, and it can also be swept for with binoculars – but **never sweep around unless the Sun is completely below the horizon, as there is always the danger of looking at the Sun by mistake**. Telescopes fitted with setting circles are essential when the Sun is in view.

Very little can be seen on Mercury's surface, and almost all our knowledge of the topography has been gained by one probe, Mariner 10. However, observers can carry out a useful piece of research by timing the date of dichotomy, or exact half-phase. With Venus, dichotomy is always early during evening elongations and late during morning elongations; this is generally known as

Schröter's effect (a term coined by the Editor of this *Yearbook* many years ago, since the effect was first noted by the pioneer lunar and planetary observer Johann Hieronymus Schröter in the 1790s). Presumably it is due to Venus' dense atmosphere. Mercury, however, has no appreciable atmosphere. Whether or not there is a similar phase effect has never been decided; this month gives a good opportunity for checking on it.

TWO CENTENARIES. April 1990 marks the centenary of the births of two notable astronomers. Walter Grotrian was born at Aachen on April 21, 1890; he joined the staff of the Potsdam Astrophysical Observatory in 1922, and became its Director in 1951. He was primarily a solar spectroscopist, who made important contributions to studies of the Sun; with A. Unsöld he edited the German journal *Zeitschrift für Astrophysik*. He died in March 1954. Also born in April 1890 was Paul McNally, SJ, who was Director of the Georgetown Observatory from 1928 to 1948; like Grotrian, he was interested mainly in the Sun, and took part in various expeditions. He was also very interested in calendar reform. He died in April 1955.

MAY

Full Moon: May 9   *New Moon:* May 24

MERCURY attains its greatest western elongation (25°) on the last day of the month. For observers in the latitudes of the British Isles the lengthening duration of twilight renders observation imposs-ible, but nearer the Equator and in the more populous regions of the Southern Hemisphere Mercury can be seen as a morning object low above the eastern horizon at the time of beginning of morning civil twilight, during the second half of the month.

VENUS is still a splendid object in the early mornings, with a magnitude of −4.0. Observers in northern temperate latitudes will find, however, that it is only visible for a short while before dawn, low above the eastern horizon.

MARS continues to be visible in the south-eastern quadrant of the sky in the early mornings. During the month Mars moves from Aquarius into Pisces as will be seen by reference to Figure 8, given with the monthly notes for December. In this area, south of the square of Pegasus, Mars is the brightest object, with a magnitude of +0.7.

JUPITER continues to be visible as a prominent object in the western sky in the evenings.

SATURN continues to be visible in the south-eastern sky in the mornings, magnitude +0.4.

PLUTO is at opposition on May 7, in Serpens, but is not visible to the naked eye (magnitude +14). It is then 4,290 million kilometres from the Earth.

THE ATMOSPHERE OF PLUTO. Pluto, discovered by Clyde Tombaugh in 1930, is the enigma of the Solar System. It is only 2,445 km in diameter – smaller than the Moon – and has an attendant, Charon, with a diameter of 1,199 km; the rotation period of Pluto, 6.3 days, is the same as the revolution period of Charon, so that to an observer on Pluto it would seem as though Charon were 'locked' in the sky. Apparently Pluto has a coating, partially at least, of methane frost, while with Charon the coating is of water frost.

An atmosphere round Pluto had been reported, but was confirmed in striking fashion by Australian and New Zealand astronomers in June 1988. On the 9th of that month Pluto occulted a star. The event was visible from Australia and from North Island of New Zealand, and extensive observations were made. Colonel Arthur Page, one of Australia's foremost amateurs who produces work of full professional value, was particularly fortunate inasmuch as the occultation was central from his observatory at Mount Tamborine, in Queensland.

It was found that before and after occultation, the star's light was dimmed; from Auckland Observatory the star was completely blotted out for 100 seconds, leaving only the faint light of Pluto. From the Carter Observatory in Wellington, on the 'edge' of the track, Graham Blow and John Priestley found that the star's light diminished by 75 per cent. From Mount John in South Island of New Zealand, where there was no actual occultation by the planet, Alan Gilmore and his colleagues found a diminution of 20 per cent.

Combining these and other observations, it was established that Pluto has a very extensive atmosphere, extending to over 600 km from the surface. There was also evidence of a strange extension on the southern side of the planet, perhaps due to the gravitational pull of Charon.

Since Pluto is so small, and is composed presumably of a mixture of rock and ice, how can it retain such an atmosphere? Of course the atmospheric density is low, but it is certainly greater than anyone had expected. The reason must be the very low temperature of about −220 degrees Centigrade. Methane, which must make up the bulk of the atmosphere, will sublime at this temperature (that is to say, change directly from a solid into a gas without going through a liquid stage). But remember, too, that as Pluto reaches its perihelion, it is closer-in to the Sun than Neptune can ever be. From now on it will move outward again, and in 1999 its distance from the Sun will be the same as the mean distance of Neptune, though the

17-degree inclination of Pluto's orbit means that no collision or even near-encounter can occur.

When Pluto moves out, the temperature will fall even further, and it is quite likely that the methane in the atmosphere will condense out. When it is near aphelion, as it will be in 124 years' time, Pluto may have virtually no atmosphere at all.

It is a curious state of affairs. We cannot even be sure that Pluto is worthy of planetary status; it may be a 'maverick' asteroid-pair, or it may be in a class of its own. There have been suggestions that it is an ex-satellite of Neptune, though this idea seems to have been out of favour recently.

Unfortunately, none of the current space-probes will go anywhere near Pluto and Charon, and we may have to wait for a long time before we find out much more about them. However, Pluto, at opposition this month, is bright enough at magnitude 14 to be seen with telescopes of the size used by many amateurs, and it is certainly worth finding.

JUNE

Full Moon: June 8   *New Moon:* June 22

Solstice: June 21

MERCURY will continue to be visible as a morning object in the eastern sky before dawn for the first three weeks of the month, though not to observers in northern temperate latitudes.

VENUS is a splendid object in the morning skies visible above the eastern horizon before dawn. Observers in the British Isles will find that it is gradually visible for a little longer each morning as the month progresses. This effect is caused by its northward movement in declination which more than offsets the fact that Venus is slowly moving in towards the Sun.

MARS continues to be visible in the mornings. Its magnitude is +0.4 and, of course, the reddish tint of the planet helps to identify it, especially as it is in a region where it is obviously the brightest object.

JUPITER is still a prominent object in the western sky in the early evenings though it is noticeably drawing in towards the Sun and observation will become impossible before the end of the month.

SATURN is a morning object, magnitude +0.2, and now visible well before midnight even to Northern Hemisphere observers.

URANUS is at opposition on June 29, in Sagittarius, when it will then be 2,752 million kilometres from the Earth. The planet is only just visible to the naked eye under good conditions having a magnitude of +5.6. Viewed through a telescope it exhibits a disk 4 seconds of arc in diameter with a slightly greenish tinge.

THREE PLANETS IN SAGITTARIUS. In mid-1990 the three outer giants, Saturn, Uranus and Neptune, are all in Sagittarius. This means that from Britain and the northern United States they are inconveniently low down, but southern observers see them high in the sky. During June Uranus, at magnitude 5.6 this month, lies roughly between the two naked-eye stars Mu and Lambda Sagittarii, and is easy to recognize with a small telescope, since it shows a small but unmistakeable disk. It comes to opposition on June 29.

Neptune, at magnitude 7.7, is below naked-eye visibility, and in small telescopes it looks very like a star.

R CORONÆ. One of the most remarkable variable stars in the sky, R Coronæ Borealis, is well on view this month. It lies in the 'bowl' of the Northern Crown, and is usually on the fringe of naked-eye visibility; binoculars show it easily, and there is a good comparison star of magnitude 6.6 in the bowl. However, R Coronæ is subject to sudden fades, when the magnitude has been known to fall as low as 15. It is thought that these fades are due to carbon accumulating in the star's atmosphere; in fact, it hides itself behind a veil of soot!

There was a fade in 1988. Whether another will be in progress by the time this *Yearbook* is published cannot be predicted. Look at the Crown with binoculars; if you see only one obvious star, you may be sure that R Coronæ is undergoing one of its minima. After a fade, it may be many weeks or even several months before the star regains its full light. It is not unique, but it is true that R Coronæ stars are among the rarest of all types of variable stars.

JULY

EARTH is at aphelion (farthest from the Sun) on July 4 at a distance of 152 million kilometres.

MERCURY is at superior conjunction on July 2. It becomes visible as an evening object for the second half of the month, though not for observers in northern temperate latitudes where the continuing lengthy twilight makes observation impossible. Observers further south should refer to Figure 6, given with the notes for August.

VENUS continues to be visible as a brilliant object in the morning skies before sunrise. From the latitudes of the British Isles the planet is visible low in the east before 03h.

MARS continues to be visible as a morning object, magnitude +0.1, in the south-eastern quadrant of the sky.

JUPITER remains too close to the Sun for observation throughout the month as it passes through conjunction on July 15. In fact the conjunction is so close that the planet passes directly behind the Sun's disk.

SATURN is at opposition on July 14, magnitude +0.1, and is therefore available for observation throughout the night. Its path amongst the stars is shown in Figure 5. At opposition Saturn is 1,345 million kilometres from the Earth.

NEPTUNE is at opposition on July 5 in Sagittarius, at a distance from the Earth of 4,367 kilometres. Having a magnitude of +7.9 it is too faint to be seen with the unaided eye.

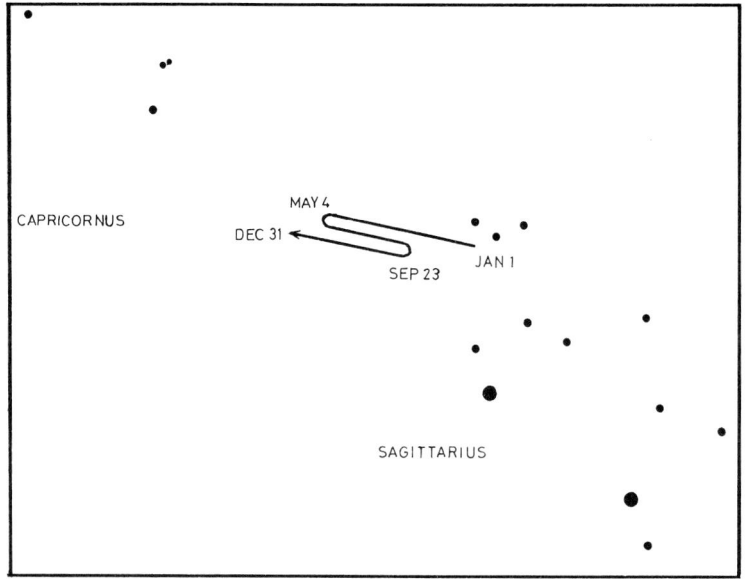

Figure 5. The path of Saturn.

There is a total eclipse of the Sun on July 22. Details are given on page 110.

HYPERION. By 1800 seven satellites of Saturn were known. Two more were added before 1900; Hyperion, discovered by Bond in 1848, and Phœbe, discovered by W. H. Pickering in 1898 – the first satellite discovery to be made photographically. All the fainter satellites have been found since 1978.

Phœbe is the furthest-out of the satellites; it has retrograde motion, and is probably asteroidal. Unfortunately it was not closely passed by either Voyager, so that we know little about its surface except that it is darkish; it has a rotation period of less than 10 hours, though it takes 550.4 days to complete one orbit. Hyperion, however, is in a different category, and it too has provided astronomers with plenty of problems.

It is a fair-sized body. Voyager measurements show that it measures $360 \times 280 \times 226$ kilometres, and its shape has been likened to that of a hamburger. Theoretically, its longest axis ought to point toward Saturn, but this is not the case. Moreover, the rotation period has been aptly described as 'chaotic'. It is at present about 13 days, though the orbital period is over 21 days; the rotation is therefore not synchronous – though Iapetus, much further out, does have the conventional synchronous rotation period (79 Earth-days).

Hyperion should be large enough to be fairly spherical, but this is not true. It has even been suggested that it is part of a larger body which suffered disruption, but in this case we are entitled to ask what happened to the rest of it.

Hyperion is less reflective than the icy inner satellites. Presumably ice is a major constituent, but there may be a 'dirty' layer covering wide areas (and it is worth noting that the much larger Iapetus has one bright hemisphere and one dark hemisphere). On Hyperion, the Voyager results show several craters, which have diameters of up to 120 kilometres; the four largest have been named Bahloo, Helios, Jarilo and Meri. There is also a ridge, which has been named Bond-Lassell. It was Bond who first identified Hyperion, but William Lassell, the celebrated British amateur, found it independently very shortly afterwards.

In 1904 W. H. Pickering, discoverer of Phœbe, reported another satellite, moving between the orbits of Hyperion and Titan. It was even given a name – Themis – but it was not confirmed, and seems not to exist; presumably Pickering mistook a dim star for a satellite.

At magnitude 14.2, Hyperion is not an easy object. Its mean angular maximum elongation from Saturn is about 225 seconds of arc.

VEGA. Apart from Arcturus, Vega is the brightest star in the northern hemisphere of the sky; its magnitude is only slightly below zero. From Britain it is near the zenith during summer evenings. July is the best time for southern observers to find it; it is well above the horizon from South Africa and Australia, though from the extreme south of New Zealand it barely rises. It was the first star found by IRAS, the Infra-Red Astronomical Satellite of 1983, to have a 'huge infrared excess' indicative of cool, possibly planet-forming material, though from this it would be rash to conclude that Vega is the centre of a true planetary system.

AUGUST

Full Moon: August 6   *New Moon:* August 20

MERCURY reaches its greatest eastern elongation (27°) on August 11 and is visible as an evening object, though not to observers as far north as the British Isles. For observers in the Southern Hemisphere this is the most suitable evening apparition of the year. For observers in latitude S.35°, Figure 6 shows the changes in azimuth and altitude of Mercury on successive evenings when the Sun is 6° below the horizon. At this time of year and in this latitude this condition, known as the end of evening civil twilight, occurs about 30 minutes after sunset. The changes in the brightness of the planet are roughly indicated by the sizes of the circles which mark its position at five-day intervals. It will be noticed that Mercury is at its brightest before it reaches greatest eastern elongation, Mercury's magnitude is 0 at the beginning of the month and it has faded to +2 by the end.

VENUS continues to be visible as a splendid object in the morning skies before dawn.

MARS is a morning object, magnitude −0.2 , and visible low in the south-eastern sky before local midnight. By the end of the month Mars is just south of the Pleiades and west of the Hyades.

JUPITER is rather too close to the Sun for observation for about the first week of August but then becomes visible for a short while in the mornings low in the eastern sky before twilight inhibits observation. Observers who know the constellation boundaries well (unlike the author!) will notice that Jupiter has now moved into Cancer.

SATURN, just past opposition, is visible for the greater part of the night, though by the end of the month observers in the latitudes of

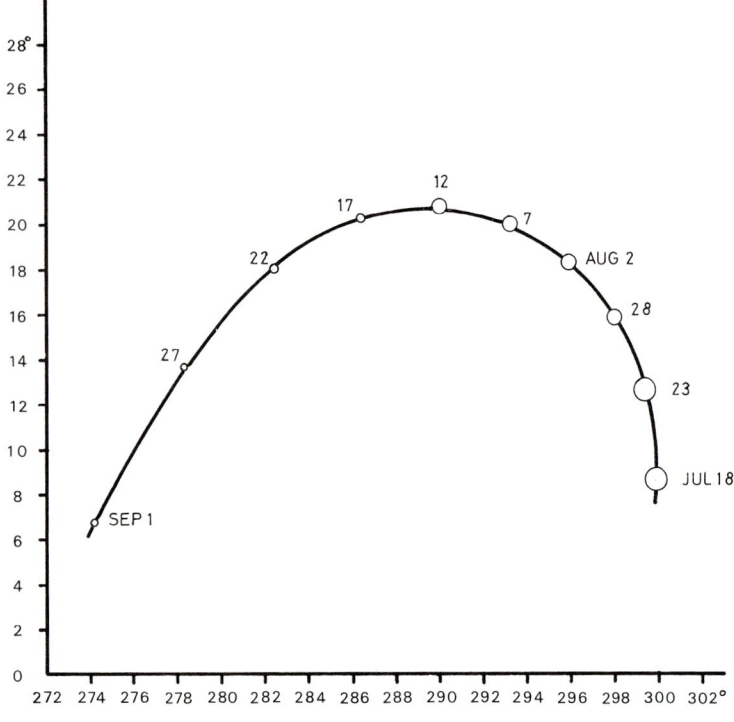

Figure 6. Evening apparition of Mercury for latitude S.35°.

the British Isles will note that it has become too low in the south-western sky to be seen after midnight.

There will be a partial eclipse of the Moon on August 6.

MAGELLAN. A year ago, Voyager 2 by-passed Neptune, thereby ending the present phase of exploration of the outer planets. This August, another American probe is scheduled to reach the neighbourhood of Venus. It has been named Magellan, after the famous Portuguese explorer.

Magellan has not had a smooth history. It began as VOIR, the Venus Orbiting Imaging Radar programme, in the 1980s. Cost-cutting ruled out VOIR in its original form; it became the Venus

Radar Mapper, now re-christened Magellan. The total cost was reduced to about 530 million dollars – which sounds a great deal until it is compared with a national armaments budget, or the amount spent yearly on tobacco or chewing-gum!

Magellan's scientific equipment was steadily cut down, and every effort was made to economize; for example, the large 3.7-metre antenna was originally built as a Voyager spare. The first idea was to use the Centaur launcher, which would have meant that the probe would have reached Venus in only six months, but instead it had to be sent on its way by a much cheaper solid-fuelled rocket, and put into an elliptical orbit round Venus instead of the ideal circular path. It will enter a polar orbit inclined to the planet's equator by 86 degrees; the revolution period will be 3.15 hours, and the minimum distance from Venus' surface a mere 250 kilometres.

Magellan, in its final form, was launched in April 1989. Its sole mission is to map the surface of Venus by radar, using the SAR equipment (synthetic-aperture radar). It will do so in strips across the surface, from 17 to 28 km broad, and the resolution may be down to 250 metres.

The first phase – one Venus year, of 243 Earth-days – will be spent in mapping; about 90 per cent of the surface will be covered. However, it is likely that there will be enough power left for additional manœuvres, and the procedure should be repeated at once; as we know, many satellites and probes go on working for much longer than their official life-times. (Remember IUE, the International Ultra-Violet Explorer!)

Parts of Venus have been mapped by the US Pioneer and the various Soviet Veneras, but much remains to be done. For example, we are still not sure whether or not active vulcanism is going on; most astronomers believe so, but positive proof has yet to be obtained, and surveys of the volcanic regions ought to help. We also want to know more about the tesseræ, or regions with intersecting ridges and grooves, and we want to find out the nature of the radar-bright flows which are thought to be due to basaltic lava. There are also craters, some of which are widely attributed to impact even though the thick atmosphere of Venus acts as an effective shield against all but meteoroids of at least kilometre size.

As we know, much of Venus is covered by a vast, rolling plain; there are two large upland areas, Aphrodite Terra (about the size of Africa) and Ishtar Terra (comparable in size with Australia), as well as high mountains and obviously volcanic regions such as Beta

Regio, with its two large shield volcanoes Rhea Mons and Theia Mons.

Venus and the Earth are almost twins in size and mass. Conditions on their surfaces are very different, and this must be due to Venus' lesser distance from the Sun. It seems strange now to recall that only thirty years ago, before the flight of the first successful Venus probe (Mariner 2), many people believed that as a prospective colony Venus was more inviting than Mars.

SEPTEMBER

Full Moon: September 5   *New Moon:* September 19

Equinox: September 23

MERCURY may still be glimpsed as a difficult evening object by observers in the tropics but only for the first two days of the month. It is at inferior conjunction on September 8 and reaches greatest western elongation sixteen days later. Thus it is a morning object for the second half of the month. For observers in the Northern Hemisphere this is the most suitable morning apparition of the year. Figure 7 shows, for observers in latitude N.52°, the changes in azimuth and altitude of Mercury on successive mornings when the Sun is 6° below the horizon. At this time of year and in this latitude this condition, known as the beginning of morning civil twilight occurs about 35 minutes before sunrise. The changes in brightness of Mercury are indicated approximately by the size of the circles which mark its position at five-day intervals. It will be noticed that Mercury is brightest after it reaches greatest western elongation. On September 16 its magnitude is $+1.7$ and two weeks later it is -0.9.

VENUS remains a brilliant object low in the eastern sky before dawn, magnitude -3.9. However, it is getting noticeably closer to the Sun. The relative positions of Venus and Mercury during the second half of the month are shown in Figure 7.

MARS is a conspicuous object in the sky, visible from the late evening onwards. During the month it increases in brightness by half a magnitude, from -0.4 to -0.9.

JUPITER is a splendid object in the eastern sky in the mornings before sunrise.

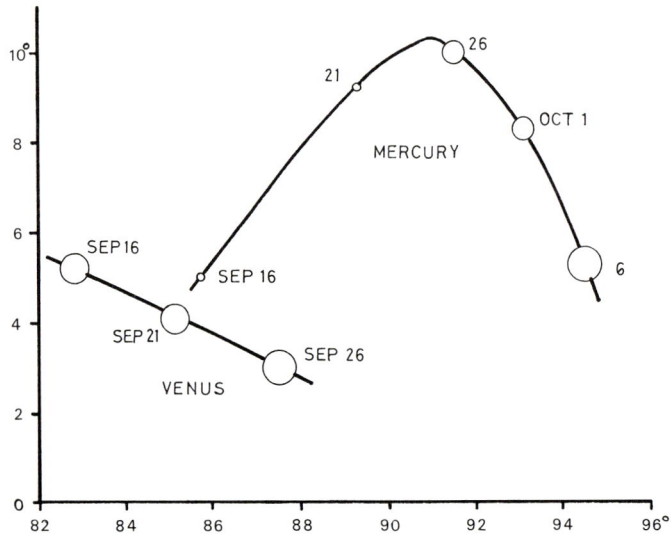

Figure 7. Morning apparition of Mercury for latitude N.52°.

SATURN is an evening object in Sagittarius.

S ANDROMEDÆ. One question often asked is: 'How far can you see with the naked eye?' The answer is: 'Over two million light-years' – because we now know that the Andromeda Spiral, Messier 31, is 2.2 million light-years away, and on a clear night it is easily visible without optical aid as a dim, misty patch. Obviously this is easier for Northern Hemisphere observers than for Southern Hemisphere dwellers, because Andromeda rises higher in the sky, but it is by no means a severe test for anyone with keen eyes. In the year 964 Al-Sufi described it as 'a little cloud', though, strangely, it was not mentioned by Tycho Brahe. The first telescopic observation of it seems to have been made in December 1612 by Simon Marius, who described it as 'like the flame of a candle seen through horn'.

In 1885 a supernova flared up in Messier 31, and reached the fringe of naked-eye visibility; it is remembered as S Andromedæ. At the time the nature of Messier 31 itself was not known, and it was usually believed to be contained in our own Galaxy (a view which persisted right up to the work of Hubble in 1923). After a while S Andromedæ faded away, and was lost in the background glow of the Spiral, but in 1989 its remnant was identified by a team led by R. A. Fesen, using a CCD with the 4-metre reflector at Kitt Peak in Arizona. The image, taken in iron light, showed the remnant as a dark spot in absorption.

We now know that the supernova was of Type I, involving the complete destruction of the white dwarf component of a binary system. The expanding cloud is iron-rich; the débris now has a diameter of about one light-year, which means that the expansion rate since 1885 has been from 4,000 to 5,000 kilometres per second. It must block out the light from up to a million stars in the Spiral.

Strangely, S Andromedæ was formerly cited in favour of the 'local' nature of Messier 31. It was reasoned that if it were a normal nova, it would have to be fairly near in order to reach the sixth magnitude; nobody then had any idea that it was an outburst of incomparably greater violence than an ordinary nova.

No supernovæ have since been seen in Messier 31, and in fact the only other supernova to have reached naked-eye visibility since then has been SN 1987A in the Large Cloud of Magellan, a mere 170,000 light-years from us. Unlike S Andromedæ, the Cloud supernova was of Type II, produced by the collapse of a single, massive star – in this case a blue supergiant.

OCTOBER

Full Moon: October 4   *New Moon:* October 18

Summer Time in Great Britain and Northern Ireland ends on October 28.

MERCURY continues to be visible as a morning object, low above the eastern horizon before dawn, but only for the first week of the month.

It is too close to the Sun to be observed for the remainder of the month, passing through superior conjunction on October 22.

VENUS is still a splendid morning object very low in the eastern sky shortly before dawn. Before the middle of the month it is too close to the Sun to be observed by those in northern temperate latitudes. Observers further south will be lucky to glimpse the planet at all, even during the first few days of October.

MARS is still brightening noticeably and this month shows the largest increase in brightness, from −0.9 to −1.6. Its rate of motion eastwards decreases as it passes north of the Hyades, until it reaches its first stationary point on October 20.

JUPITER is a brilliant morning object, magnitude −2.1. By the end of October observers in northern temperate latitudes will be able to see Jupiter rising in the east before midnight. Jupiter is in Gemini and its path amongst the stars is shown in Figure 1, given with the monthly notes for January.

SATURN continues to be visible as an evening object, magnitude +0.5.

THE BRITISH ASTRONOMICAL ASSOCIATION. This month marks a very important centenary. The British Astronomical Association was formed in 1890, and had its first meeting in October of that year. It was held at the Hall of the Society of Arts, John Adam Street, Adelphi, on October 24. Captain W. Noble was elected President; the Editor was E. W. Maunder, who had been largely responsible for the formation of the B.A.A.

The first discussion involved the name! Originally it had been 'The British Astronomical Society', but this was thought to be too close to the Royal Astronomical Society, and the change was made immediately. It is interesting now to look back at the first Section Directors, appointed at that meeting: Miss Brown (Sun), T. G. Elger (Moon), W. R. Waugh (Jupiter), W. S. Franks (Star Colours), J. E. Gore (Variable Stars), K. J. Tarrant (Double Stars), D. Booth (Meteors), and T. E. Espin (Spectroscopic and Photographic). Of these, the Star Colour Section died long ago, and the Double Star Section was never a success.

Ever since 1890 the B.A.A. has had an observational record second to none, and its publications, notably its Journal and Memoirs, are mines of information. Of course, other Sections have been added, and the present membership is over 2,000 all told.

The B.A.A. has always been primarily an amateur organization, though with a strong professional element; there have been many professional Presidents, notably Sir Harold Spencer Jones at the time when he was Astronomer Royal.

This year there are many celebrations in honour of the Centenary. No doubt the next hundred years of the Association's existence will be as fruitful as the past century has been.

ENCKE'S COMET. This month sees the return to perihelion of the periodical comet with the shortest known period: Encke's. It was first seen in January 1786 by Pierre Méchain, who, apart from Messier, was the most successful comet-hunter of the time; it was then of the fifth magnitude. It was under observation for less than a week, and was not seen again until November 1795, when it was found by Caroline Herschel. Again it was lost; it was picked up again in October 1805 by Pons, and developed a 3-degree tail, with a magnitude which became as bright as 4. J. F. Encke calculated a period of 12 years, but it was not recovered until November 1818, again by Pons. It was then that Encke identified it with the previous comets, and found a period of 3.3 years; he predicted a return for 1822, and in June of that year it was duly recovered, by C. L.

Rümker in Australia, close to the position given by Encke. Since then it has been seen at every return except that of 1944, when it was badly placed and most astronomers had other things on their minds!

The aphelion distance is little over 4 astronomical units, and the comet can now be followed all round its orbit with the aid of modern instruments. The brightest return seems to have been that of 1829, when the magnitude reached 3.5. The tail is seldom very marked, and has never exceeded the 3-degree length seen in 1805.

It is likely that the comet is now fainter than it used to be; this is inevitable, since it loses material at every return to perihelion. There have been suggestions that it will 'die' some time during the next century, but this is by no means certain, and astronomers would certainly regret the loss of a comet which they have come to regard as an old friend.

NOVEMBER

Full Moon: November 2   *New Moon:* November 17

MERCURY is slowly emerging from the evening twilight and for the second half of the month is visible as an evening object, though not for observers as far north as the British Isles.

VENUS reaches superior conjunction on the first day of the month and then moves very slowly east of the Sun. Even by the end of November it is still only 7° from the Sun, too close for observation.

MARS reaches opposition on November 27 and is visible throughout the night: it is moving slowly westwards in Taurus. Its magnitude at opposition is −2.0, brighter than any star or planet in the sky until Jupiter rises – this planet is only a few tenths of a magnitude brighter than Mars, though noticeably different in colour. Because of the eccentricity of the orbit of Mars the times of closest approach and of opposition will only rarely coincide. On this occasion closest approach occurs nearly 8 days before opposition. At closest approach Mars is 77 million kilometres from the Earth, almost a million kilometres closer than when at opposition.

JUPITER continues to be visible as a brilliant morning object, magnitude −2.2.

SATURN is still an evening object in the south-western sky, but poorly placed for observers in the British Isles because of its low altitude.

MARS. The 1988 opposition of Mars was the most favourable of recent years, because the planet was at opposition and near perihelion at the same time. In 1990 things will not be so favourable, but will still be better than they will be again for the rest of the century.

The high northern declination means, however, that observers from countries such as South Africa, Australia and New Zealand will have to cope with a rather low altitude; British and North American observers will have much the best of matters.

At the time of opposition, it will be late summer in the southern hemisphere of Mars, and winter in the north. The tilt of the axis means that we will have good views of both polar regions, whereas in 1988 it was the southern pole which was presented. Whether or not there will be any major dust-storms remains to be seen; during 1988 they were absent, so that the surface features remained sharp and clear-cut throughout the whole of the apparition.

THE LEONIDS. In the early morning of November 17, the Leonid meteor shower is at maximum. In the past there have been magnificent displays, but the last of these was in 1966, and we do not seriously expect another before 1998 or 1999; however, it is always worth keeping a watch – the Leonids are quite unpredictable.

ACHERNAR. For Southern Hemisphere observers, the bright star Achernar is near the zenith during November evenings. With a magnitude of 0.46, it is in fact the ninth brightest star in the entire sky. It is 85 light-years away, with a spectral type of B5 and a luminosity 780 times that of the Sun according to the authoritative Cambridge catalogue.

Achernar's declination is −57 degrees, which means that it cannot be seen anywhere north of latitude N.33° on the Earth. It is the leader of Eridanus, the River, a long, sprawling constellation which stretches from Achernar to Kursa (or Beta Eridani) close to Rigel in Orion. Achernar has been called 'the End of the River', but this has also been associated with the third-magnitude star Acamar (Theta Eridani), which has a declination of −40 degrees.

It is worth noting that the south celestial pole lies about midway between Achernar and the Southern Cross, so that during November evenings, when Achernar is high, Australian or South African observers will find that the Cross grazes the horizon – though from much of New Zealand the Cross is always well above the southern horizon.

DECEMBER

Full Moon: December 2 and 31   *New Moon:* December 17

Solstice: December 22

MERCURY, being so far south of the Equator, is always so low above the horizon after sunset that observers in the British Isles will be unable to detect the planet even though it is at greatest eastern elongation (21°) on December 6. For observers further south Mercury will continue to be visible as an evening object for the first half of the month. Mercury passes through inferior conjunction on December 24.

VENUS is too close to the Sun for observation at first, but gradually moves outwards and so becomes visible in the evening skies low above the south-western horizon, but only for a short while after sunset. Observers in northern temperate latitudes will have difficulty in seeing the planet until about ten days before the end of the year, because Venus is so far south of the Equator.

MARS is only just past opposition so it continues to be visible for most of the hours of darkness. It will fade noticeably during December, the magnitude changing from −1.9 to −1.0. Figure 8 shows the path of Mars relative to the stellar background for the later months of the year.

JUPITER does not actually come to opposition until the end of next month but it is already clearly visible in the night sky from mid-evening onwards. Its magnitude is −2.4 and it is moving slowly retrograde in Cancer.

SATURN is now coming towards the end of its evening apparition so that the planet is only visible for a short while in the south-western

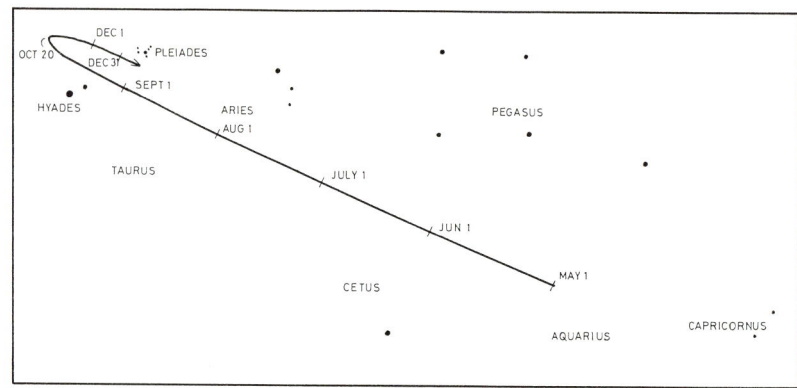

Figure 8. The path of Mars.

sky in the early evening. Observers in northern temperate latitudes will have lost it in the evening twilight before the end of December.

THE CELESTIAL CRAB. During December, the brilliant planet Jupiter is moving among the stars of Cancer, the Crab. Cancer is one of the least conspicuous of the Zodiacal constellations; it has only two stars above the third magnitude, Beta (3.5) and Delta (3.9). The outline is not unlike that of a very dim and ghostly Orion. The constellation is not hard to find; it lies between Regulus in Leo to one side, and the Twins (Castor and Pollux) to the other.

The most interesting objects in the constellation are the two open clusters, M.44 (Præsepe) and M.67 (see diagram). Præsepe, sometimes nicknamed the Beehive, is the brighter of the two, and is an easy naked-eye object against a dark sky; it is flanked by the stars known as the Aselli or 'Asses', Delta (Asellus Australis) and Gamma (Asellus Borealis, magnitude 4.7). Apart from the Pleiades, it is probably the finest cluster in the Northern Hemisphere of the sky. M.67, in the same binocular field as Alpha Cancri or Acubens (magnitude 4.2) it is on the fringe of naked-eye visibility, and is very easy to see with binoculars; it is notable because it seems to be among the very oldest of all known open clusters.

Note also the very red semi-regular variable X Cancri, near Delta. It is of spectral type N, and since its range is between magnitudes 5.5 and 7.5 it is always within binocular range. There is a rough period of about 195 days.

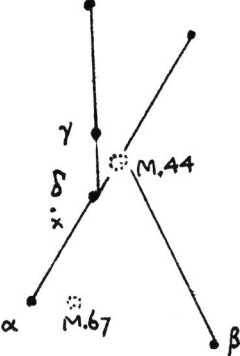

Cancer, the Celestial Crab

THE CHRISTMAS STAR. As Venus begins to emerge into the evening sky toward the very end of the year, the usual question will be asked: 'Can it have been the Star of Bethlehem?' The answer is a very emphatic 'No', because Venus was a well-known object in the time of Christ, and its appearance would have caused no excitement whatsoever. The familiar suggestion that the Star of Bethlehem might have been due to a close conjunction of two planets seems improbable for the same reason (and would have lasted for a considerable time). No nova or supernova at this period was recorded by contemporary astronomers, and Halley's Comet came back much too early. There is in fact no scientific explanation for the Biblical story – but one thing is certain: Venus was not responsible.

Eclipses in 1990

In 1990 there will be four eclipses, two of the Sun, and two of the Moon.

1. An annular eclipse of the Sun on January 26 is visible as a partial eclipse from South Island, New Zealand, the eastern part of the South Pacific Ocean, the Southern Ocean, Antarctica except the extreme east, South America except the north-western part and the western part of the South Atlantic Ocean. The eclipse begins at 17^h 13^m and ends at 21^h 48^m. The annular phase begins in Antarctica at 18^h 55^m and ends at 20^h 06^m in the South Atlantic Ocean. The maximum duration of the annular phase is 2^m 03^s.

2. A total eclipse of the Moon on February 9 is visible from northwest Alaska, the Arctic regions, North Pacific Ocean, Australasia, Asia, the Indian Ocean, part of Antarctica, Africa, Europe including the British Isles, the north-east Atlantic Ocean, Iceland and Greenland. The eclipse begins at 17^h 30^m and ends at 20^h 54^m. Totality lasts from 18^h 51^m to 19^h 33^m.

3. A total eclipse of the Sun on July 22. The path of totality begins on the southern coast of Finland, passes along the north coast of Asia and across the extreme north-eastern part of Asia and ends in the Pacific Ocean between the Hawaiian Islands and North America. The partial phase is visible from Scandinavia except the southwest, north of Greenland, north of Asia, the Arctic regions, northwestern part of North America, the Hawaiian Islands and the North Pacific Ocean. The eclipse begins at 00^h 40^m and ends at 05^h 24^m, the total phase begins at 01^h 53^m and ends at 04^h 11^m. The maximum duration of totality is 2^m 33^s.

4. A partial eclipse of the Moon on August 6 is visible from southwest Alaska, the Pacific Ocean except the extreme eastern part, Antarctica, Australasia, the south and eastern parts of Asia and the eastern part of the Indian Ocean. The eclipse begins at 12^h 43^m and ends at 15^h 39^m. The time of maximum eclipse is 14^h 11^m when 0.68 of the Moon's diameter is obscured.

Occultations in 1990

In the course of its journey round the sky each month, the Moon passes in front of all the stars in its path and the timing of these occultations is useful in fixing the position and motion of the Moon. The Moon's orbit is tilted at more than five degrees to the ecliptic, but it is not fixed in space. It twists steadily westwards at a rate of about twenty degrees a year, a complete revolution taking 18.6 years, during which time all the stars that lie within about six and a half degrees of the ecliptic will be occulted. The occultations of any one star continue month after month until the Moon's path has twisted away from the star but only a few of these occultations will be visible at any one place in hours of darkness.

There are nine occultations of planets in 1990, one each of Mercury, Venus and Mars, and three each of Jupiter and Saturn. None of these is visible from Great Britain.

Only four first-magnitude stars are near enough to the ecliptic to be occulted by the Moon; these are Regulus, Aldebaran, Spica, and Antares. Of these four only Antares is occulted (on twelve occasions) in 1990.

Predictions of these occultations are made on a world-wide basis for all stars down to magnitude 7.5, and sometimes even fainter. Lunar occultations of radio sources are also of interest – remember the first quasar, 3C.273, was discovered as the result of an occultation.

Recently occultations of stars by planets (including minor planets) and satellites have aroused considerable attention.

The exact timing of such events gives valuable information about positions, sizes, orbits, atmospheres and sometimes of the presence of satellites. The discovery of the rings of Uranus in 1977 was the unexpected result of the observations made of a predicted occultation of a faint star by Uranus. The duration of an occultation by a satellite or minor planet is quite small (usually of the order of a minute or less). If observations are made from a number of stations it is possible to deduce the size of the planet.

The observations need to be made either photoelectrically or visually. The high accuracy of the method can readily be appreciated when one realizes that even a stop-watch timing accurate to $0^s.1$ is, on average, equivalent to an accuracy of about 1 kilometre in the chord measured across the minor planet.

Comets in 1990

The appearance of a bright comet is a rare event which can never be predicted in advance, because this class of object travels round the Sun in an enormous orbit with a period which may well be many thousands of years. There are therefore no records of the previous appearances of these bodies, and we are unable to follow their wanderings through space.

Comets of short period, on the other hand, return at regular intervals, and attract a good deal of attention from astronomers. Unfortunately they are all faint objects, and are recovered and followed by photographic methods using large telescopes. Most of these short-period comets travel in orbits of small inclination which reach out to the orbit of Jupiter, and it is this planet which is mainly responsible for the severe perturbations which many of these comets undergo. Unlike the planets, comets may be seen in any part of the sky, but since their distances from the Earth are similar to those of the planets their apparent movements in the sky are also somewhat similar, and some of them may be followed for long periods of time.

The following periodic comets are expected to return to perihelion in 1990:

Comet	Year of discovery	Period (years)	Predicted date of perihelion 1990
Tuttle–Giacobini–Kresak	1858	5.5	Feb. 8
Sanguin	1977	12.5	Apr. 2
Schwassmann–Wachmann (3)	1930	5.3	May 19
Russell (3)	1983	7.5	May 29
Peters–Hartley	1846	8.1	June 21
Russell (4)	1984	6.6	July 7
Tritton	1978	6.3	July 8
Honda–Mrkos–Pajdusakova	1948	5.3	Sept. 12
Encke	1785	3.3	Oct. 28
Kearns–Kwee	1963	9.0	Nov. 5
Johnson	1949	7.0	Nov. 18
Wild (2)	1978	6.4	Dec. 16
Taylor	1915	7.0	Dec. 28

Minor Planets in 1990

Although many thousands of minor planets (asteroids) are known to exist, only 3,000 of these have well-determined orbits and are listed in the catalogues. Most of these orbits lie entirely between the orbits of Mars and Jupiter. All of these bodies are quite small, and even the largest, Ceres, is believed to be only about 1,000 kilometres in diameter. Thus, they are necessarily faint objects, and although a number of them are within the reach of a small telescope few of them ever reach any considerable brightness. The first four that were discovered are named Ceres, Pallas, Juno and Vesta. Actually the largest four minor planets are Ceres, Pallas, Vesta and Hygiea. Vesta can occasionally be seen with the naked eye and this is most likely to occur when an opposition occurs near June, since Vesta would then be at perihelion. In 1990 Juno will be at opposition on May 8 (magnitude 10.1) and Vesta on November 15 (magnitude 6.6).

A vigorous campaign for observing the occultations of stars by the minor planets has produced improved values for the dimensions of some of them, as well as the suggestion that some of these planets may be accompanied by satellites. Many of these observations have been made photoelectrically. However, amateur observers have found renewed interest in the minor planets since it has been shown that their visual timings of an occultation of a star by a minor planet are accurate enough to lead to reliable determinations of diameter. As a consequence many groups of observers all over the world are now organizing themselves for expeditions should the predicted track of such an occultation cross their country.

In 1984 the British Astronomical Association formed a special Asteroid and Remote Planets Section.

Meteors in 1990

Meteors ('shooting stars') may be seen on any clear moonless night, but on certain nights of the year their number increases noticeably. This occurs when the Earth chances to intersect a concentration of meteoric dust moving in an orbit around the Sun. If the dust is well spread out in space, the resulting shower of meteors may last for several days. The word 'shower' must not be misinterpreted – only on very rare occasions have the meteors been so numerous as to resemble snowflakes falling.

If the meteor tracks are marked on a star map and traced backwards, a number of them will be found to intersect in a point (or a small area of the sky) which marks the radiant of the shower. This gives the direction from which the meteors have come.

The following table gives some of the more easily observed showers with their radiants; interference by moonlight is shown by the letter M.

Limiting dates	Shower	Maximum	R.A. Dec.	
Jan. 1–6	Quadrantids	Jan. 3	$15^h28^m+50°$	
April 20–22	Lyrids	April 22	$18^h08^m+32°$	
May 1–8	Aquarids	May 5	$22^h20^m+00°$	
June 17–26	Ophiuchids	June 19	$17^h20^m-20°$	
July 15–Aug. 15	Delta Aquarids	July 29	$22^h36^m-17°$	M
July 15–Aug. 20	Pisces Australids	July 31	$22^h40^m-30°$	M
July 15–Aug. 25	Capricornids	Aug. 2	$20^h36^m-10°$	M
July 27–Aug. 17	Perseids	Aug. 12	$3^h04^m+58°$	M
Oct. 15–25	Orionids	Oct. 22	$6^h24^m+15°$	
Oct. 26–Nov. 16	Taurids	Nov. 3	$3^h44^m+14°$	M
Nov. 15–19	Leonids	Nov. 17	$10^h08^m+22°$	
Dec. 9–14	Geminids	Dec. 13	$7^h28^m+32°$	
Dec. 17–24	Ursids	Dec. 22	$14^h28^m+78°$	

M = moonlight interferes

Some Events in 1991

ECLIPSES

There will be three eclipses, two of the Sun and one of the Moon.

January 15–16: annular eclipse of the Sun – Australasia, Antarctica.
July 11: total eclipse of the Sun – the Americas.
December 21: partial eclipse of the Moon – Iceland, Greenland, the Americas, Australasia, Asia, N. Scandinavia.

THE PLANETS

Mercury may be seen more easily from northern latitudes in the evenings about the time of greatest eastern elongation (March 27) and in the mornings around greatest western elongation (September 7). In the Southern Hemisphere the dates are May 12 (morning) and November 19 (evening).

Venus is visible in the evenings until July and in the mornings from September to the end of the year.

Mars is visible in the evenings until September.

Jupiter is at opposition on January 29.

Saturn is at opposition on July 27.

Uranus is at opposition on July 4.

Neptune is at opposition on July 8.

Pluto is at opposition on May 10.

A Run on the PTI

DAVID ALLEN

The half-mile walk from the quarters is particularly pleasant this afternoon. A heavy cloud cover gives protection from a sun that can be so scorching even now, quite early in the summer. I stroll along the straight avenue planted with conifers and native casuarinas, while crested pigeons take to their wings with a chatter of complaint. So many varieties of birds flock to this avenue that every tree issues a different song. Now a flock of thornbills twitters in the low branches; now comes the pleasant tone of a red-rumped parrot.

How much more enjoyable is the walk than just eleven hours earlier when I had dragged myself with throbbing head from a feverish sleep to relieve one of my collaborators in the control tower. Two of our team were ill, one had been working since 8 a.m. and the fourth had just driven 350 km from Sydney and was also in need of sleep. After five hours in bed I was arguably the fittest of the four, and since the rules require two people in the tower whenever observing takes place, I could justify sleeping no longer.

The huge dish always seems ghostly at night. It stands at the south end of the avenue, dimly lit save for two flashing red beacons to warn low-flying aircraft. From afar it looks tiny; only as your footsteps carry you closer do you come to appreciate the full 64 metres of its diameter, a gargantuan upturned bowl of girders and mesh. By night as a man approaches the tower on which it sits he cannot but feel dwarfed to insignificance. This is the Parkes radio telescope, one of the hallowed sites to which all who dabble in radio astronomy should undertake one pilgrimage. It is also the only memorial still standing to the engineering genius of Barnes Wallis, a man we normally associate with the dam busters, but who also designed the interleaved spiral girders of the Parkes dish.

When I arrived last night, and hauled myself up the two floors to the control room, I found the place subdued. Normally this would be a sign of things going wrong, as well they might in the highly complicated mode of observing we had elected to use. But no: all was

Figure 1. A distant view of the Parkes radio telescope.

working remarkably well; only tiredness could be blamed for the lassitude. On my arrival a brief debate ensued over which of the two men should take his turn to collapse into bed. The debate was short-lived. Ray, who had driven from Sydney, was definitely in worse shape, even though Euan had been on the go now for 17 hours; gratefully Ray departed.

I quickly assessed how things were going. The log book documented the steady garnering of data, showing that we were averaging about 11 minutes per source, of which nearly 9 were occupied in the measurement and the remainder covered the ponderous slewing of the great dish. Indeed, not only of this dish, but also of an even larger structure 275 km to our south at Tidbinbilla (Tid).

We were using the Parkes–Tidbinbilla interferometer, the PTI, which is proudly described as the world's largest real-time interferometer. A grandiose title, perhaps, but a superb instrument that I am privileged to use largely because Ray himself had been its engineer. He had noted that NASA's satellite tracking dish, just south of Canberra at Tidbinbilla, sometimes stands idle. By installing some clever software Ray worked out how to link it to Parkes so that the two could be used as an interferometer. The Tidbinbilla dish was also 64 metres in diameter, and recently it has been enlarged to 70 metres in readiness to receive the weak signals from the Voyager spacecraft in its encounter with Neptune. We were thus using two of the world's largest radio telescopes as a giant binocular.

Anyone who has lost the use of one eye, even for a short period, will be aware of the advantages of binocular vision. For radio astronomy the gains of an interferometer over single-dish observing are much greater. The radio receiver determines something the eye cannot, namely the phase of the incoming wave. If the crest of a wave reaches both dishes at the same time, the signal adds; if a crest at one dish coincides with a trough at the other then the signal is cancelled out. Therefore the interferometer can tell if the source of the waves is tilted relative to a chosen direction by an angle equal to that subtended by half a wavelength at the separation of the dishes.

In the more complicated case of an extended source moving across the sky, an interferometer records a signal only if there is sufficient energy coming from within a very small angular spread on the sky. In the case of the PTI, at the wavelength of 13 cm we were using, we would detect a radio source only if it were no larger than one-tenth of a second of arc in diameter, about the apparent size of a pin head at a distance of 2 km, and a much smaller angle than optical telescopes are capable of resolving because they are restricted by the turbulence of the Earth's atmosphere.

When using the PTI, the observers operate the Parkes telescope and the controlling programmes, whilst a telescope operator looks after the Tidbinbilla end of things. We communicate to Tid by means of a computer terminal, its screen showing the messages that have passed back and forth. At Tidbinbilla the telescope operator gets from the screen not only the text we type in, but also the coordinates of the source we require next. Whenever we change source, he has to enter these in and press a button to send back an acknowledgement. After a while the part of the screen devoted to his messages is filled with a series of lines alternating the words 'On our way' and 'Source change acknowledged by Tidbinbilla', whereas our last missives remain, frozen in time like a recording of a telephone conversation in which one hears only the voice at one end. When I had arrived last night the screen read:

> sorry wrong source
> this'n
> yep
> ok. thanks for everything.
> ok

It is an impersonal way to communicate. I don't even know the name of the Tid operator, and merely assume him to be male.

The 'real-time' part of the glossy name describes the other valuable aspect of the PTI. Usually when radio telescopes are linked together over such distances the data are recorded on tape and only put together weeks later to produce the full information. With the PTI, data from Tidbinbilla pours directly into Parkes as we sit here, and the computer combines them as we watch. If we observe a strong source we can tell immediately that we are detecting it; a weak source is usually apparent, and only the very faintest must await a more complete analysis before they show. Even that analysis can be undertaken within a few minutes of the observation. We can therefore make sensible decisions as we observe, knowing roughly how the observing plans are working out.

I am brought back to this afternoon by the scolding of a magpie from one of the trees. A dusky apostle bird skulks disdainfully under a low branch, while overhead a pair of galahs swoops, their pink undersides set against a sky of deeper hue than the soft grey of their backs and upper wings. Many are the disparaging comments made about galahs, yet to me they are one of the loveliest sights of inland Australia. Oft have I admired great flocks of them wheeling in perfect synchrony so that they switch from grey to pink or back to grey all in a single instant. As I near the end of the avenue I leave behind the trees and their contents, and begin to wonder what Ray has found in the more detailed analysis of last night's data.

At the foot of the tower I pass an unsightly collection of temporary huts whose origins remind me of the part Parkes has played in several spacecraft operations. The telescope was instrumental in receiving signals from the first Apollo astronauts to land on the Moon, and more recently from Giotto's tour through the heart of Comet Halley and Voyager's passage of the Uranus system. It will again be receiving signals when Voyager reaches Neptune, an encounter that will have taken place by the time you read this.

The tower is a three-storey cylinder topped by the dish and its drive mechanisms. The ground floor does not tempt the casual visitor, being the home of an air-conditioning plant together with general storage. I go there only for its one vital facility, but on every such visit I recall the occasion when the hind leg of a frog protruded from under the rim of the toilet bowl, and I wonder whether he/she is still in residence.

Figure 2. From the catwalk below the dish one has an unusual view of the control tower beneath. The top row of windows illuminates the observing room.

The first floor is largely taken up by a computer, but there is a small office/kitchenette locally known as Rest Point, a pun on Tasmania's famous Wrest Point casino. On a long night the kitchenette and its contents are as vital to survival as the facilities lower down. A new consignment of cream biscuits has arrived, I note; they won't last long.

The second floor is taken up by the control room. A tapered central column leaves a doughnut-shaped room of ample proportions. Except at a doorway, the column is girdled by desks and tables, and the outer wall of the tower houses huge racks of electronics, computer terminals, printers, and graph plotters, and a few more desks. It is clean and comfortably carpeted, and a pleasant place to work.

There is no observing taking place as I arrive, because the Tidbinbilla dish is not currently available. None the less, the room is a hive of activity. Not only are two of my collaborators beavering away examining our data and dealing with other tasks, but some of the resident staff are here developing software, testing electronics and the like.

There is work to do before we take an early dinner. The observing programme will be updated taking account of the results from last night. Moreover, there are complex calibration corrections to be applied to the data. Plots are steadily streaming off one of the machines: jagged black lines that summarize each observation more succinctly than a string of numbers. Here I find one with a weak source detected, a spike that rises above the remaining wiggles like a tall blade of grass. The next shows no prominent spike and adds to the pile of non-detections. For every ten minutes of observation, millions of numbers have been boiled down virtually to a simple vote: detection, yes or no.

Tid will start at 6 p.m., so dinner is at five. Now we drive to the quarters, as time is valuable. I miss the sound of the birds. We eat well and return, replete, at 5.45 to make the first impersonal contact with Tid and give everything a final check over. Then both dishes are sent to a well-established calibration source, Pks 1921-293, as a final verification. The real business can now begin.

As last night, the staccato pace of 10-minute observations begins. Data pour out of the system. When we have a batch of observations, we run them through the initial analysis programme and study the results. I have been building up a simple bar chart on a sheet of paper, showing how the number of detections depends on the type of galaxy we are observing, for this is the thrust of our research.

Before I describe more fully what we are doing, I must warn you that I am going to lead you away from the known, reliable facts you will find in books, and present instead some of the medley of ideas, suspicions and hunches that sometimes can lead a scientist to useful discoveries. This is science in the raw, science that has as much chance of being totally wrong as it has of making another break-through.

Our interest is not in ordinary galaxies, but in those with unusual activity evidenced by heated gas near their centres. Only a trace of gas is involved, but it must be illuminated by some intense ultra-violet or X-radiation. The stars that usually inhabit the hearts of galaxies are too cool to provide the heating we see in some

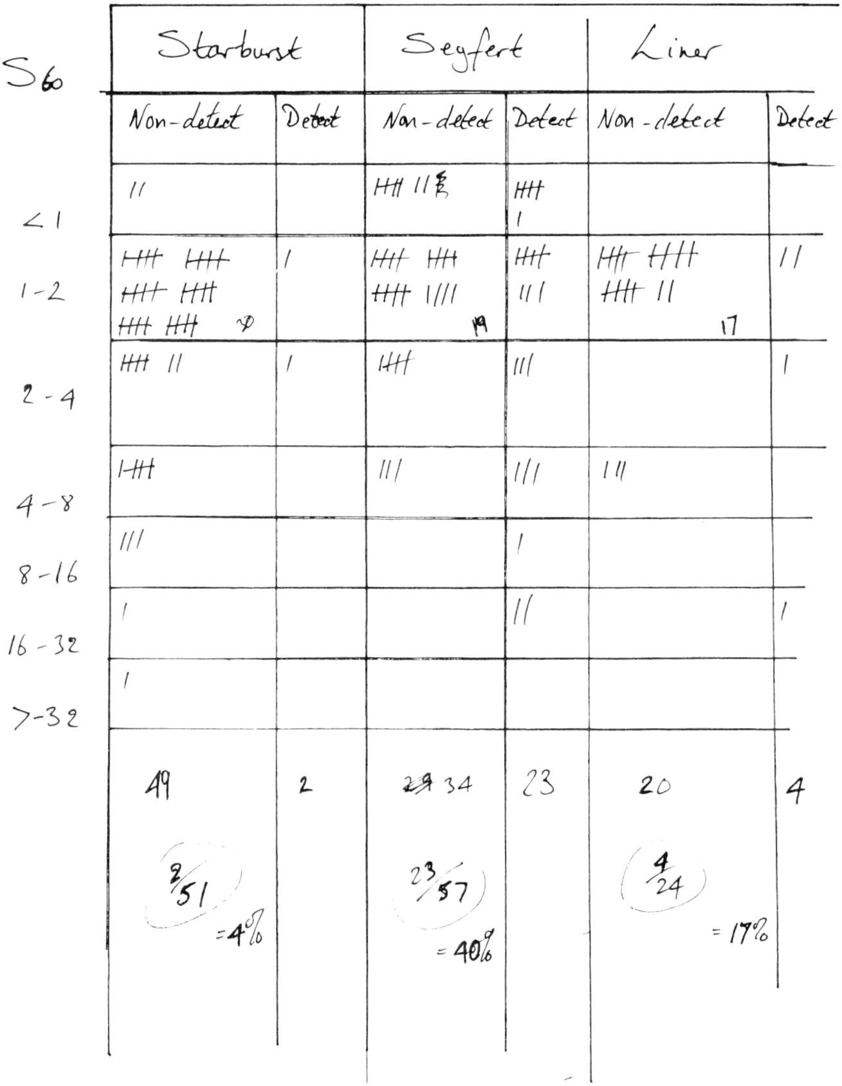

S_{60}	Starburst Non-detect	Starburst Detect	Seyfert Non-detect	Seyfert Detect	Liner Non-detect	Liner Detect
< 1	//		HH //	HHt /		
1 - 2	HHt HHt HHt HHt HHt HHt ∿	/	HHt HHt HHt //// ₁₉	HHt // /	HHt HHt HHt // ₁₇	//
2 - 4	HHt //	/	HHt	///		/
4 - 8	HHt		///	///	/ //	
8 - 16	///			/		
16 - 32	/			//		/
> 32	/					
	49	2	29 34	23	20	4
	$\frac{2}{51}$ = 4%		$\frac{23}{57}$ = 40%		$\frac{4}{24}$ = 17%	

Figure 3. On a pad of note paper I have kept track of the detections according to galaxy type (starburst or Seyfert) and their brightness in the IRAS infrared survey.

examples. Two explanations have been proposed for the extra source of ultraviolet energy in active galaxies: unusually hot stars, or miniature quasars.

Consider first the hot stars. Regions in which hot stars have formed are dotted throughout the outer parts of our and other galaxies; the Orion Nebula is such a place. The young stars there provide ultraviolet radiation aplenty to excite the gas to glow. But hot stars use their supplies of hydrogen quickly, and within a few million years all will vanish, dramatically, as supernovæ. The stars that populate the centres of normal galaxies were formed thousands of millions of years ago, using up all the available gas in the process. Only the cool ones now remain. If we are to invoke hot stars to explain the heating of gas in galactic cores, we must believe that a new injection of gas has been made near their centres in sufficient amount to have given birth to millions of new stars in the last million years. There is, in fact, clear evidence that this has happened in a few cases, often (perhaps always) as a result of the collision of two galaxies. If one of the colliding galaxies has gas in its outer parts, as do most spiral galaxies, then that gas will tend to fall to the centre of the single galaxy that results from the collision. There follows a rapid episode of star formation, liberating ultraviolet energy copiously. Galaxies in which this phenomenon is thought to be taking place are called starburst galaxies.

The mini-quasar hypothesis also has strong observational support in many cases. One of the defining features of a quasar is its optical spectrum, which is quite distinctive. The same spectrum shows in many of the active galaxies, for example in Messier 77, a galaxy in the constellation Cetius. We see quasar-like objects with a huge range of luminosities; the most extreme give off about ten million million times the output of the Sun, whilst the smallest are a million times weaker, and there is little reason to doubt that still weaker examples are hidden by the surrounding stars in other galaxies. Galaxies whose nuclei are quasar-like but lie at the weaker end of the range of luminosity are known as Seyfert galaxies, after an American astronomer who first drew attention to them nearly fifty years ago.

If you were to ask professional astronomers unofficially what they think is going on in quasars, at least 80 per cent, and possibly 95 per cent, would tell you that a black hole is swallowing gas and stars. If you were to ask them to put this belief formally in print, most would hedge. We have not satisfactorily proved that black holes lie at the centres of any galaxies; indeed it isn't clear that we have absolutely proved that black holes exist anywhere. However, there is now general acceptance that these queer phenomena, predicted by Einstein's theory of relativity, can form and probably have done so in our and other galaxies.

Black holes are a state of matter from which nothing can escape – not even light – but they retain one important influence on the outside world. They have gravity. Once a black hole has formed, it draws to itself and consumes any material lying nearby. In so doing an enormous amount of energy is liberated, hundreds of times greater, mass for mass, than is available to stars when they convert hydrogen to helium. A small black hole therefore grows larger, and a black hole is capable of liberating energy in proportion to its mass if there is enough material to feed on. We believe that quasars involve rampant black holes, gorging on the hearts of galaxies, because we can conceive of no other source of energy great enough to explain them. In a luminous quasar, the black hole must have grown to a mass one hundred million times that of the Sun. Since it may have started out in the range 10–100 times the solar mass, such a monster has done a fair amount of feeding, and the galaxies in question must be somewhat hollowed out in their innermost regions.

The neat division of active galaxies into starbursts and Seyferts has been questioned of late. Many argue that there is no obvious reason why black holes couldn't exist in galaxies undergoing starbursts, for even the largest of black holes could not consume all the gas that might be dumped near the centre of a galaxy before it had a chance to turn into stars. More pressing, the discovery by the IRAS satellite that some galaxies emit copious amounts of infrared radiation has questioned whether the starburst picture is really adequate.

IRAS galaxies have been mentioned in the last two issues of the *Yearbook*, and I make no apology for raising them again here. For they are an important discovery. In a nutshell, IRAS found many galaxies which give off as much energy as quasars, but do so at long infrared wavelengths because the visible and ultraviolet light has been intercepted by dusty clouds of gas which process it into infrared radiation. Yet the optical spectra of these galaxies are by no means Seyferts, and most resemble starbursts. If we argue that what looks like a starburst galaxy really is a starburst galaxy, then we are lumbered with explaining how on Earth an ordinary galaxy can be a hundred times as luminous as an ordinary galaxy. If this sounds like something from Lewis Carroll, then I should explain that it is just about possible to make things fit by turning lots of gas into really big stars, for they put out all their energy quickly and for a brief period could indeed produce the required energy. But how a galaxy can condense stars so quickly is not at all obvious.

A much simpler explanation, therefore, is to argue that quasars are lurking at the centres of the IRAS galaxies: that they are disguised Seyferts. This suggestion, with its various pros and cons, is one of the central issues at present, and many astronomers are busily trying to find ways to test it.

A keyboard gives three electronic peeps. Above my head a disconcerting rumble begins. I have observed at Parkes often enough not to be alarmed when a thousand-odd tons of steel begins to move just above the ceiling, but the memory is fresh in everyone's mind of the 300-foot telescope at the National Radio Astronomy Observatory at Green Bank, West Virginia, that fell down unexpectedly only a few days ago. More electronic bleeps tell us that the Tidbinbilla dish is also on the move. The motors above my head slow and stop, and the screen in front of me flashes the message 'Waiting for Tidbinbilla to acknowledge'. Three more tones from the PTI screen follow, the message vanishes, and a series of bleeps of a different pitch emanates from the rack of electronics off to my right. The calibration is being fired for a few seconds. Then that too is silenced, and we begin 516 seconds of measurement on another galaxy.

My thoughts pass through the door in the central cone of the control room, a route I have followed on several occasions. Up a spiral stair one enters the m. e. room, a grey, steel enclosure which houses a small optical telescope. This is the master equatorial, a telescope that has not seen the sky for thirty odd years, yet which serves a vital purpose whenever observing is in progress. The m. e. always points faithfully at the object being observed. It is a relatively simple matter to point a small telescope, protected from the elements, with great accuracy. Because the telescope is equatorially mounted it also tracks the object unerringly. The same is not true of a huge radio telescope, pivoted in altitude and azimuth so that it must motor in both directions at changing rates to follow an object across the sky, and subject to enormous wind pressures. So the little m. e. directs the big dish above it, keeping it always pointing correctly, by a very simple trick.

On the underside of the dish is a mirror. The m. e. sends a light beam up to that mirror and receives the reflection. If the dish is squinting away from the desired direction, the light beam will return to the wrong spot. Its location is sensed by light cells that send signals to the motors and cause the dish to be driven into alignment. You may recognize this scheme by several names. In optical telescopes it would be called an autoguider, and would ensure that a big telescope continued to track a star correctly. In engineering parlance it is a servo mechanism.

Above the m. e. room one enters the structure of the dish itself. Here the rooms tip with the dish, and they are fashioned like the companionways of a ship. It is possible to continue to the outside world this way, immediately under the dish where a catwalk leads almost to the outer edge and a small ladder takes you through a hole on to the reflecting surface at the base of one of the tripod legs that supports the feed cabin. Those who prefer fresh air can reach the same catwalk by an external route from the control room.

A tiny lift, just big enough for two good friends, takes the explorer up the tripod leg, and at the top it is possible for the moderately agile to enter a fair-sized room at the focus of the dish. Here sits the receiver, the device that converts the incident radio waves into measurable electric currents. Receivers used to be little more than bits of wire, much as you find in a domestic radio set. Now they are much more sophisticated. Computer-designed cones transfer the signal into huge cryostats, where electronic components are cooled to temperatures as low as a few degrees above absolute zero. Ever bigger pieces of equipment are being fitted into the feed cabin, so that there are times when there is little room for a human being, particularly since several receivers can be installed at any one time. Up there, in the darkness where my thoughts have drifted, the little receiver is merrily grabbing waves of radio emission that left the galaxy we were observing one thousand million years ago.

On one of the plethora of computer screens a pattern of small crosses is displayed, one for every two seconds we have been observing this galaxy. Periodically the screen is redrawn to bring it up to date with the steady flow of data. Usually the crosses are entirely random, looking as though they had been spilled from a bag, and this is the case now. Sometimes, however, as on the previous galaxy, the eye could pick out a tendency for the crosses to line up in a broad band that slants across the screen. If the band disappears off the top of the screen, it comes in again at the bottom.

Figure 4. Three steel legs rise from the dish of the Parkes telescope. The nearest includes at its base the small lift that rises to the antenna cabin at the top.

Such a band indicates that a source has been detected, and the degree of tilting is a measure of how close the source lies to the position we specified when we set the telescopes. In the log book I note 'no obvious source'. An hour or so hence, when we have another batch ready for processing, we will run the data through the analysis programme and see whether a weak source is indeed present. Then will I fill in a few more ticks on a notepad.

The ticks are my record of progress. Through them I keep track of the objects we do and do not detect. On one side of the page I record the galaxies classified from optical spectra as starbursts; on the other side the Seyferts and their cousins the Liners, which are thought to be even weaker versions of Seyferts. To give us more feel for what the data show, I have further subdivided the galaxies according to their intensity as measured by IRAS. Naïvely we think we are more likely to detect strong IRAS sources than weak ones.

The data so far present a telling story. We have detected only a couple of starburst galaxies out of scores. But we are finding almost one third of the Seyferts. If this rate keeps up till the end of the observing run we will have absolutely incontrovertible evidence that the two types of galaxy really are different. We will be able to say that Seyfert nuclei are hidden in very few of the starburst galaxies, and that most of them – even those of extreme luminosity – are powered by something else. Presumably that something else is, indeed, stars.

This may seem a very small point, and indeed we are not about to write to the Nobel prize committee, or issue press releases claiming what brilliant science we have done. This is but one piece in a huge jigsaw, a mote of research which will have to be incorporated in the blanket of knowledge we slowly weave and which the theorists strive ever to explain. A feeling of satisfaction is shared by those of us in the control room that this is a fairly significant portion of the pattern, a bit that both theorists and posterity itself will long admire. But then one always has an inflated view of the importance of one's own piece of research, especially when the data are first examined. Time is a great leveller.

Indeed, this very control room has witnessed much more important activities. It was here, in 1963, that an observation was made of the strong radio source 3C 273 when the Moon passed in front of it. At the time interferometers like the PTI didn't exist, and radio astronomy had the myope's view of the sky. Locating radio sources with sufficient precision to be able to find an optical counterpart for further study was, at the time, extremely difficult but absolutely fundamental to further progress. Radio astronomy had gone as far as it could go without the help of other wavebands. The occultation provided such a means, for the signal was cut off exactly as the limb of the Moon crossed the source. All that was needed was to time exactly when this happened on two occasions – as the source vanished behind the Moon and again as it reappeared – to pinpoint the place exactly.

The observation was successful, despite the fact that the source was very low in the sky, and 3C 273 was identified with a faint star-like object that had a bizarre optical spectrum. Thus were discovered the quasars.

Now, a quarter century later, I am seated in the same room using the same telescope (with a little help from its friend down south) to search for quasars hidden from the view of optical telescopes, and by mapping the sky in far greater detail than optical telescopes can attain. No longer can radio astronomy be called myopic.

The story of the discovery of 3C 273 reminds me of one anecdote that I just have time to tell before we process the next set of data. It concerns Miller Goss, now the director of the VLA radio telescope in New Mexico, and I don't think Miller would mind my telling it.

A decade ago Miller spent some time in Australia, and used the Parkes telescope extensively. So often, in fact, that he became qualified to operate it himself. On one occasion he wanted to observe very low in the sky, and started moving the dish down. Something went wrong and the dish began to run away. This was no problem, as the control software would stop it before it hit the ground. Or so Miller believed. But a bug had appeared in the software, and the telescope sailed through the limit without a pause.

No worry, though. There is a mechanical switch that is activated if the telescope does somehow get past the software limit, and this cuts the power to the motors. At least, it should . . . only it hadn't been tested for a while and had jammed open. A final limit fortunately existed, the one last barrier that prevented the dish from ramming into the ground and being bent or damaged. Somewhere atop the tower a ring of metal sits, so sited as to engage the structure of the dish in a robust place and absolutely stop it in its tracks. It does so with a mighty shudder, and those in the control room are probably scared half to death. But it is infallible.

All this happened in a few seconds, and Miller could do no more than brace himself for the imminent juddering stop. It just happened, however, that he had turned the dish to the exact azimuth at which 3C 273 had sat for that occultation measurement. To allow the experiment to take place, a small portion of the metal ring had been machined away.

Miller was awarded a 'leather medal' for excavating a trench outside the tower and deforming the Parkes dish. Luckily the damage was slight and easily rectified, and Miller took away a souvenir that I expect he displays to this day in his office at the VLA.

Time for a Change of Time in the UK?

GORDON TAYLOR

Suggestions have been made that the UK should permanently keep a time of G.M.T.+1 hour. In other words the standard time of the UK would become Central European Time. We have, of course, been keeping G.M.T.+1h as Summer Time between March and October for many years. What would the change mean?

First, let us take a look at the basic structure of our timekeeping system. Currently our standard time is Greenwich Mean Time (G.M.T.), also known as Western European Time (W.E.T.). Standard times are referred to a particular meridian of longitude and were chosen so that when the mean Sun crosses that meridian the standard time is 12h, thus dividing the hours of daylight into two almost equal parts. Thus

		degrees
Greenwich Mean Time (G.M.T.) is referred to longitude		0
Central European Time (C.E.T.) is referred to longitude		E.15
Eastern European Time (E.E.T.) is referred to longitude		E.30

$$C.E.T. = G.M.T. + 1h \qquad E.E.T. = G.M.T. + 2h$$

Other time zones based on multiples of 15° span the world.

As astronomers we know that, because of the equation of time, the time when the real Sun crosses a particular meridian varies slightly throughout the year, but is never more than 16 minutes different from the time of the mean Sun crossing the meridian. Noon is defined as the time when the real Sun crosses the meridian.

If we look at the map (Figure 1) we may wonder why the other countries of western Europe do not keep Western European Time. Actually, before World War II they did. When Germany invaded France it imposed C.E.T. on that country, which never reverted to W.E.T. after the cessation of hostilities. Neighbouring countries

131

have followed suit so that they now all keep C.E.T. This is quite an anomaly since C.E.T. is referred to a meridian running through Poland, east of Berlin! For once, the UK, which still keeps G.M.T., can justly claim that all the others are out of step, as will be seen in the following list of countries of the European Economic Community (EEC).

Country	Appropriate Time Zone		
	0	1h	2h
Belgium	G.M.T.		
Denmark		C.E.T.	
France	G.M.T.		
West Germany		C.E.T.	
Greece		C.E.T.	or E.E.T.
Ireland	G.M.T.		
Italy		C.E.T.	
Luxembourg	G.M.T.		
Netherlands	G.M.T.		
Portugal	G.M.T.		
Spain	G.M.T.		
United Kingdom	G.M.T.		

This list gives the standard time most appropriate to that country when its longitude is considered. In fact, before World War II this describes the situation that occurred, apart from West Germany (which did not then exist as a separate state) and the Netherlands, which, believe it or not, kept a standard time of G.M.T. −0h 19m 32.1s! Thus we see from this table and the map (Figure 1) that if a common standard time was to be kept in the EEC then it would be more logical to use G.M.T.

It is interesting to note that countries whose territory spans a considerable range of longitude have so far felt no need to use a common time. The USSR has ten different time zones and the USA has six. In addition to this variation each state of the USA may opt whether or not to keep Summer Time. Australia, which has much less variation in sunrise/sunset times than the EEC, has three time zones. If the rest of the EEC can persuade the UK, Ireland and Greece to change to C.E.T. then all twelve countries will be keeping the same time. Will the EEC be leading the way? Looking even farther ahead can we visualize a time when the whole world keeps

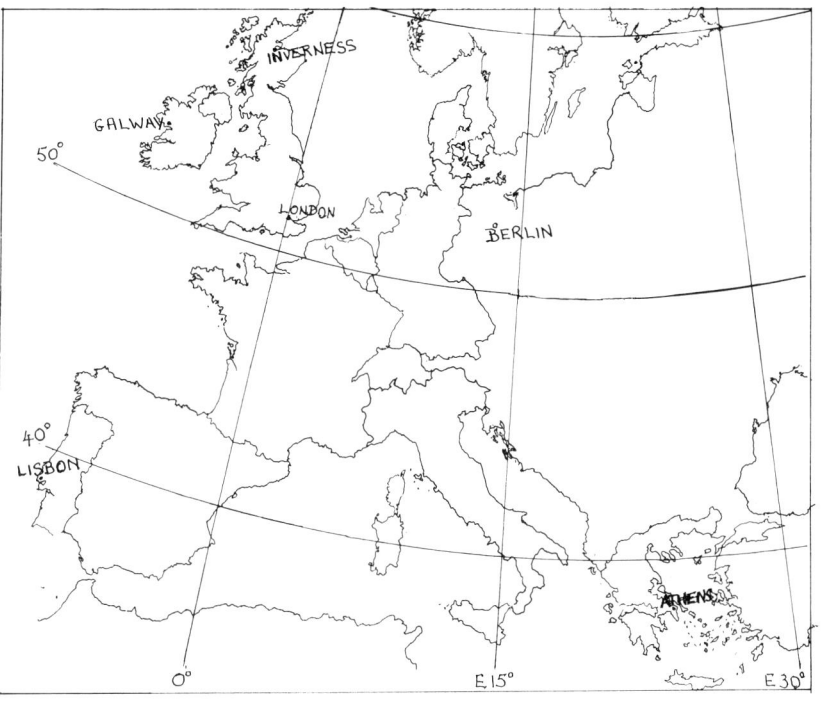

Figure 1. Map of Europe.

one standard time – even if countries in high and temperate northern latitudes still wanted to keep Summer Time?

In the UK, a time of G.M.T. + 1h was kept continuously between 1968 and 1971 as an experiment. Unfortunately an attempt was made to disguise the fact that this was keeping C.E.T. by trying to call it British Standard Time (very confusing since the standard time of Britain was still G.M.T.). Opinion was divided as to the success of the experiment. Businessmen claimed that they benefited because they were able to have longer conversations on the telephone with their continental colleagues. Therefore if we do adopt C.E.T. there will be an obvious economic benefit – to British Telecom! On the other hand, it was noted in the building trade that extra costs were involved in winter as it was necessary to provide artificial lighting just for a short period in the mornings. The experiment was abandoned.

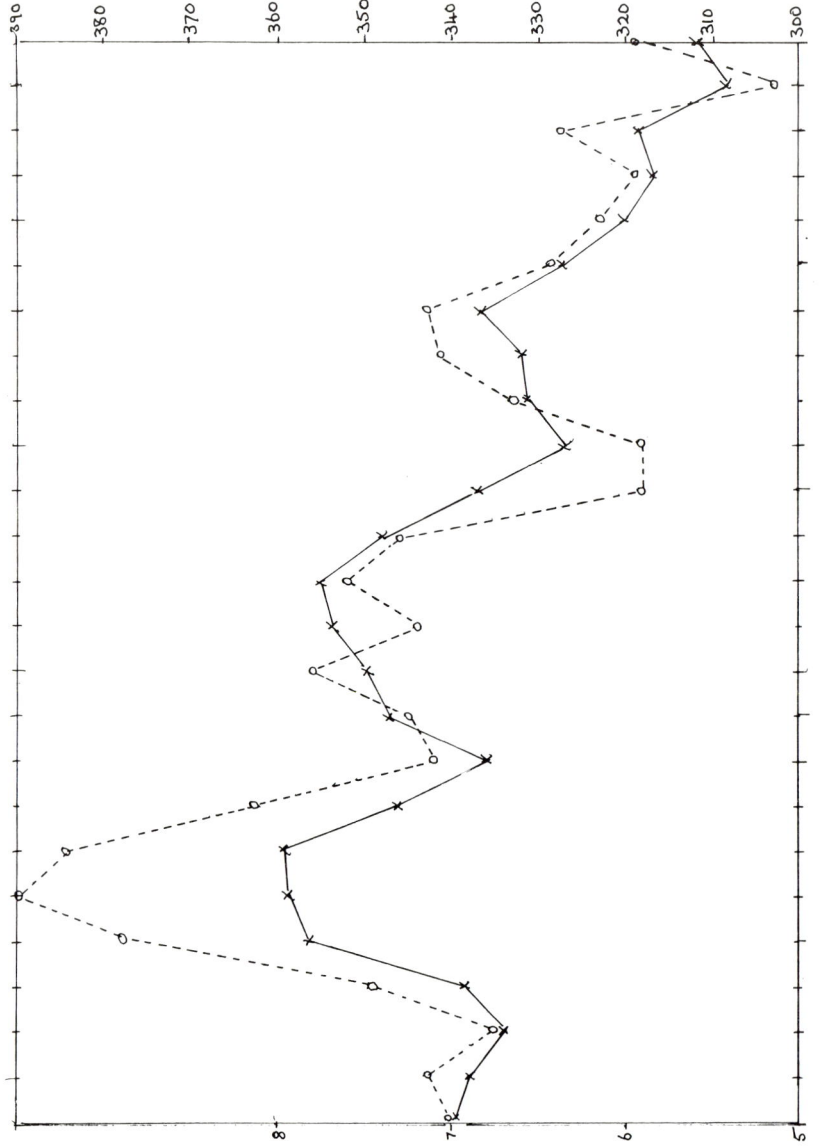

It occurs to this author that as road conditions, weatherwise, are worse just before dawn than they are just after sunset (both for temperature and visibility, on the average) that this is an additional argument for *not* moving the time of going to work/school back into the pre-dawn darkness, particularly when this would occur during the worst months of the year in the UK. Is there any supportive or detrimental evidence for this idea? The graph (Figure 2) shows the road casualty figures for a number of consecutive years, the year number increasing from left to right. The figures have been extracted directly from Whitaker's Almanack. Each point gives the value for the year. There will be two consecutive years (1969–1970) when the experiment was in force throughout the year. In 1968 the experiment started in October: it finished in 1971 October. If you think that there might have been more or less casualties during the experiment you are invited to examine the graph and see if you can detect which were the years 1968–1971. On the graph the actual numerical values of the years have been excluded so that you will not be biased. To satisfy your curiosity turn to page 157 and you will find out how to put the year numbers on the graph and see what conclusions can be drawn.

Why don't all the EEC countries keep the same time? Let's examine the data in Table 1, which gives the times (G.M.T.) on four dates of the year for a number of places in the EEC. The dates chosen are those of the equinoxes and solstices so that we can see the complete range of times throughout the year. The problem in adopting a common time lies in the wide range in both latitude and longitude. Compare the times for Athens and Inverness on December 21. The Sun rises 3h 20m earlier at Athens than at Inverness. A change to C.E.T. would mean that sunrise at Athens would be at 6h 37m and at Inverness 9h 57m. Travelling salesmen selling alarm clocks are doubtless preparing to invade Scotland!

Even if the UK abandons G.M.T. there will still be a large number of countries keeping it as their standard time, namely Ascension Island, Canary Islands, Côte d'Ivoire, Faroes, Gambia, Ghana, Guinea, Guinea Bissau, Iceland, Liberia, Madeira, Mali, Mauritania, Morocco, St Helena, Sao Tome & Principe Islands, Senegal, Sierra Leone, Togo, Tristan da Cunha and Upper Volta.

Figure 2. Road casualties in the UK. The continuous line indicates the number of killed (in thousands, read off the left-hand scale) and the dotted line indicates the number of injured (in thousands, read off the right-hand scale).

TABLE 1

Sunrise/sunset times in G.M.T.

Place	Mar. 20		June 21		Sept. 23		Dec. 21	
	rise *h m*	set *h m*	rise *h m*	set *h m*	rise *h m*	set *h m*	rise *h m*	set *h m*
Athens	4 29	16 36	3 03	17 51	4 14	16 19	5 37	15 09
Lisbon	6 41	18 48	5 12	20 05	6 25	18 33	7 51	17 19
Berlin	5 10	17 19	2 43	19 33	4 55	17 01	7 15	14 54
London	6 04	18 13	3 43	20 21	5 47	17 57	8 03	15 53
Inverness	6 20	18 30	3 18	21 19	6 03	18 14	8 57	15 32
Galway	6 40	18 49	4 08	21 08	6 23	18 33	8 49	16 20

Now what about Summer Time? The standard time of the UK has been G.M.T. since 1880 and the use of Summer Time (1 hour fast on G.M.T.) dates from 1916 when it was in operation from May 21 to October 21. Since then we have always used a form of Summer Time for part of the year. Occasionally a time 2h fast on G.M.T. was kept (1941–1945 and in 1947). The dates of change to and from Summer Time have varied considerably since 1916. From 1923 to 1939 it always started in April and this continued again, after World War II, until 1960. Since then it has always started in March (apart from the 1968–1971 experiment).

All the changes have occurred at 01h or 02h G.M.T., and always on Sundays except for the reversion to G.M.T. in 1917–1921 inclusive and the change to G.M.T.+2h in 1945, both of which occurred on Mondays.

Presumably, if the UK adopted C.E.T. as its standard time it would also be required to keep a Summer Time, like the rest of the EEC. In this case local noon in the UK would occur around 14h 00m to 14h 30m on the clock and the hottest time of day in our so-called summer would be around 16h 30m to 17h 00m. The dates when Summer Time is in operation are far more critical in Inverness than in Athens, since at the former city the times of sunrise and sunset vary through the year by nearly six hours whereas the range for the latter city is less than half as much. Sunset at Inverness, at the summer solstice, would not occur until 23h 19m, only 11 minutes later than at Galway, Ireland.

What other effects would the proposed change have? With the growth of Scottish nationalism would there be a demand for Scotland to continue to keep G.M.T.? If we abandon G.M.T. will there also be a call to abandon the Greenwich meridian as the zero longitude for all maps? Readers will have to come to their own conclusions. One thing is certain, however – the decision about any of these changes will be made by politicians, so logic is unlikely to prevail.

Would amateur astronomers benefit or not from the adoption of C.E.T. (+1h in summer)? Most of them observe in the evenings so that in the summer months normal bedtime would occur before the sky was dark enough to observe. As may be seen from Table 2, which gives the times of the various twilights, in southern England, at the summer solstice, nautical twilight does not end until after midnight.

In winter the later time of sunset would benefit only those who would otherwise get back from work too late to see the low, bright, artificial satellites which are then only visible for a short while after sunset.

TABLE 2

Twilight times

Time system →		C.E.T.	C.E.T.+1h	C.E.T.+1h	C.E.T.
Date →		Mar. 20	June 21	Sept. 23	Dec. 21
Place	*Twilight*	*h m*	*h m*	*h m*	*h m*
Greenwich	C.T.	19 48	23 10	20 32	17 34
Edinburgh	C.T.	19 51	24 06	20 36	17 13
Greenwich	N.T.	20 28	24 29	21 12	18 18
Edinburgh	N.T.	20 36	T.A.N.	21 21	18 03
Greenwich	A.T.	21 10	T.A.N.	21 53	18 59
Edinburgh	A.T.	21 23	T.A.N.	22 08	18 48

T.A.N. = Twilight lasts all night.

The table gives the times of ending of civil twilight (C.T.), nautical twilight (N.T.), and astronomical twilight (A.T.). At the end of evening civil twilight (Sun's centre 6° below the horizon) the brightest stars are visible and the horizon is still clearly defined. At the end of evening nautical twilight (Sun's centre 12° below the horizon) the sky is almost completely dark. The end of astronomical twilight (Sun's centre 18° below the horizon) marks the onset of theoretical perfect darkness.

Here are the facts. Now weigh up the evidence and make up your own minds. Should the UK keep its standard time as G.M.T. and try to persuade the EEC to adopt it, or should we change to C.E.T. – or should we agree to differ?

Eclipse Chasing

MICHAEL MAUNDER

Nobody who has seen a total solar eclipse forgets it. Many become addicted. The addiction takes many forms and can lead to a great deal of satisfaction, as well as a diminished bank balance. Because total solar eclipses are such rare events, it makes sense to be fully prepared.

The next total solar eclipse is in 1990 (see page 110 for details). The year after an even longer one will be seen in Hawaii and into Mexico. Both occur in the holiday month of July.

State of Mind

The viewer's state of mind has to be made a first priority. Why? The reason is simple. Unless there is a single-mindedness of purpose, the multitude of snags cropping up will discourage all but the most determined, and the experience of a lifetime will be missed. Nature waits for nobody. Unless you are at the right place at the right time, forget it. Start planning years, not days in advance.

Which Country?

A few hours spent in a library checking travel agents' brochures is always time well spent. It will confirm or deny the climatic conditions, *at the time of year you intend going*. Is it the rainy season, for instance? Moving along the eclipse track to another place or country can make all the difference. If 'where?' is not too important a consideration, it boils down to deciding to have a good holiday in a country you would like to go to sometime.

Some decisions on the country are made on very basic issues. Age and infirmity are something beyond control. If you have a problem, the hospital facilities then need to be adequate in case of emergency. Conversely, fit people may have important vaccine allergies. Yellow fever vaccine is based on eggs, and many cannot take it. Local regulations might require a certificate, so that country is out unless an alternative is found.

Politics and ethics are also beyond control. The old saying 'When in Rome . . .' cannot be overstated. Anyone with strong objections to a régime or conditions in a country is well advised not to go.

How to Get There

Without a doubt, the best way to get to an eclipse for the first time is to use a specialist travel agent. There are many of these advertising. Only consider alternative methods if you know the country, are an experienced traveller, or value the travel experience more. In many countries off the beaten track, the choice of hotels and accommodation is often limited, and snapped up by professionals very early on. The specialist firms also have some influence when it comes to selecting and varying the viewing site. This can be very important. A friendly police force can work wonders securing necessary travel permits or supplies.

Seasoned eclipse chasers often prefer to make their own arrangements. However, it is interesting that many such find their way to near the 'Official' groups in the end! In the case of eclipse tracks entirely over the sea, there is little option but to take what ship is on offer. The choice has to be made on cost and who is likely to be on board. Ship travel is often the answer to those with 'problems' with certain countries.

Real planning can now start in earnest.

Photography or Viewing?

The biggest single mistake to make at any time is to be too ambitious. Even experienced photographers can make mistakes, and an eclipse is simply not like an ordinary assignment. The best advice to offer newcomers is to regard photography as a bonus. Go prepared to view the eclipse with nothing more than the naked eye.

The great advantage of naked-eye viewing is that you have little or no clutter. Last minute changes of plan are easily accommodated. You also have time to stand and stare. In your mind will be the everlasting picture, and no photograph ever replaces that for vividness. Photographers should always come mentally prepared to ditch all their cherished plans (equipment sometimes!) and fallback to simple naked-eye viewing. Someone, somewhere is bound to get a picture to keep as a souvenir. Snapshots of the site and people there can also have some value later, particularly if anything unusual happens. It is worth spending a little time on the best way of seeing an eclipse. Direct viewing of the total eclipse is really the whole point of going. The dangers lie in viewing the partial phases.

Even direct viewing of the normal Sun anytime with the naked eye is fraught with hazard. Right up to a second or two either side of totality is not safe. Sun-glasses are no good, either, for more than glances.

The safest viewing method is by projecting the Sun's image on to a screen. However, this means taking extra optical equipment, and most people dodge the issue. They either ignore the partial phases, or use a filter.

The dodgers can go to extreme lengths and wear eyepatches or welder's goggles. The idea is to build up dark adaptation so that every scrap of totality is seen to best advantage. The improvement is definitely worth the effort.

The majority, which is most of us, use filters to try for the best of both options. It becomes the only option when photographing the partial phases before totality. The Sun's image must never be looked at through any form of camera viewfinder without proper filters.

Warning on Filters

This WARNING cannot be repeated often enough:

No filter is entirely safe when placed behind the main mirror or lens, just before the eye.
Never, ever, consider them, whatever the source or type.

Filters placed in front of the main optics – whether eyeball, telescope or camera – must have the right properties. Eye damage arises from radiation outside the visible spectrum.

Ultraviolet radiation (UV) is well known for its damaging effects such as sunburn. It will pass through a surprisingly large number of visually opaque filters. The main damage is to the cornea, leading to cataracts and similar irreversible optical defects.

Infrared radiation (IR) is even more prone to pass through visually opaque filters. It is heat radiation and this concentrates on the retina or in the eyeball fluid, literally cooking it. Irreversible vision loss occurs in the areas affected.

Mylar filters have become the most popular type for solar work in recent years. Mylar is an extremely tough plastic film. The type to use is 10 microns thick, coated with a very thin layer of aluminium metal, often on both sides.

Aluminium absorbs both UV and IR radiation and transmits only a small amount of visible light, mainly blue.

IMPORTANT POINTS TO WATCH WITH SOLAR FILTERS

Mylar

Mylar film must never be confused with 'Silver Paper', which is a thin sheet of aluminium metal, nor with glitters. These are similar plastic sheets but with a silver or metallic ink printed on to them. They are also becoming very common in the packing industry and are very dangerous indeed because they pass too much damaging radiation.

Only use Mylar from a reputable source.

Mylar itself is extremely tough, but the aluminium coating is not. Each piece must be inspected for pinholes. These holes transmit a lot of damaging radiation, and also cause image degradation due to halation. This is seen as lowered contrast and sharpness.

The surface must be inspected regularly for unevenness. Scuffing and other abrasions will give local highspots and pinholes. The dangers are obvious and image degradation can be appreciable. Some slight rucking or unevenness will not cause image degradation. On the contrary, some authorities positively recommend not stretching the film too tightly.

The main problem with Mylar arises from a factor which is rarely mentioned or discussed. The film is 'poled'. In other words, it has a marked polarization. You can see this by holding up a piece to sunlight. The Sun's image will often have 'wings'. These rotate as the piece is turned round. The effects really become serious with some types of camera viewfinder, particularly autofocus where circular polarization is needed. Before attempting to photograph anything important, check focus by rotating the filter and always double check with test exposures.

Silver Printed Plastics
Not to be used under any circumstances.

'Black' Colour Film
Not to be used under any circumstances.

'Black' Black and Film
Safe to use if dark enough. It gives better colours than Mylar. The

image is rarely easy to focus, and has low contrast due to halation arising from the Callier Effect.

Wratten No. 96 Filters

Optically the best in the visible range for preset or rangefinder cameras, *i.e.* where the image is not seen directly. These neutral density filters are totally transparent to infrared radiation.

Images produced through a Wratten 96 filter by direct means, such as a normal SLR viewfinder, **must not be viewed with the eye under any circumstances**.

Inconel and other metal coatings

Several commercial versions exist of this metal coating on optical glass. They are the best available and are produced to precise optical specifications. The drawback is the cost.

Preparations for Viewing

Genuine naked-eye viewing of totality needs no preparation except dark adaptation. For the partial phases, making or buying filters is all that is needed.

No eclipse is ever really complete without some form of optical aid, and binoculars must be the most popular. Any already in hand should be adequate. The choice of binoculars to buy for the first time is never easy to answer. It is hardly worthwhile buying specially; choose some to use for whatever else you do that needs them. A general purpose, lightweight modern type is unlikely to be a disaster.

Preparations for Photography

Practice and more practice with the equipment.
That is all that need be said.
And repeated.

Choice of Camera

Cost is usually the decider. However, simplicity is much more important. A totally new camera can be a recipe for disaster unless sufficient time and trouble is taken to get used to it. It has to be realized that eclipse photography is totally unpredictable and camera controls must be second nature. The controls may have to be used in what amounts to pitch darkness. Some of the more modern cameras with a range of auto functions may suit you, but they must be checked first. It is rare to find a camera which is not black these days. An old-fashioned chrome finish is a good idea when left in the Sun before and after the eclipse. Otherwise, take along a cloth cover, preferably white to reflect heat.

This advice applies equally to still and movie cameras. True movie cameras continue to give more information than video, but for how long is a matter of debate. The rate of video technology change is so rapid that there is a tendency to swap equipment before the old has been fully explored. If you do wish to use video, always give it a full testing under adverse lighting conditions where the controls cannot be seen clearly. This is particularly true when hanging a new lens on the front.

Simple still cameras, such as range-finders, are possibly the best advice for anyone only interested in filling a picture album. The simple camera's ability to cope with a wide range of lighting is legendary, but there is a most important word of warning for you here. You must bear in mind those photographers with more serious intentions.

Do not take a camera fitted with an automatic flash. Not only will you fail to record the real lighting, but you could upset a crucial picture in another camera. Worse still, you can destroy essential dark adaptation for anyone doing serious visual work.

SLR cameras remain the most popular choice, because of the wide range of lenses capable of being fitted. Make sure that you have enough adaptors and fittings when using lenses not designed by the camera maker. Check over battery consumption and buy only fresh batteries with plenty of spares, before you leave home. Battery consumption increases alarmingly in damp and very cold climates.

To re-cap, the history of eclipse trips is littered by tales of photographers failing to get pictures because they have not had a camera long enough to get used to its funny ways. Stick with an old one if in doubt. Better still, take your old camera along as well. All professional photographers carry a spare (or two!).

Choice of Lenses and Telescopes

For some strange reason, too few give this important choice the attention it deserves. Sheer image size is not the only criterion – quality has to come into it. The longer the focal length chosen, the more the quality of the *whole* system has to be taken into account. If large images are necessary, choose the longest focal length possible within your budget. Do not use a zoom – always select the best 'prime' lens you can afford. Quite often a simple glass lens will outperform a zoom for reasons outside the scope of this article. Good quality mirror telelenses (without glass close-focus facility) are best of all for colour rendition.

Having chosen a focal length, stick with it and do not add teleextenders until the particular combination has been tested fully. An extender creates at least two effects. The least serious is light loss and image degradation. A 2 × extender loses 2 stops in speed and any optical effects are also worsened and new ones added. Quite often this will be added colours, something to avoid with the subtle corona. By far the more serious effect is the worsening of camera shake. Camera shake is the main killer of image quality on eclipses. Extra-heavy lenses and attachments stuck on the camera upset its natural balance and a very sturdy tripod becomes the most essential ingredient in the whole system.

Tripods are not cheap, at least in the quality needed for long focal lengths, and money spent on a good one is often a much better investment than a fancy lens. A tripod has to be sturdy enough for the job in hand and yet not so heavy that it takes up all the aircraft luggage allowance.

Effectiveness is all important.

Many seasoned eclipse chasers do not take expensive tripods. They make a simple but rigid wooden structure and work with that. When the eclipse is over, the whole lot is left behind to give plenty of baggage allowance for souvenirs!

The tripod effectiveness must be checked for the longest focal length lens. Not just in a showroom, or even in a studio, but under the actual conditions expected at the site. Take some time to check this out. If no information is forthcoming, assume the worst. This is loose sand or springy turf. Test each lens *and* tripod combination by test exposures long before you leave and then decide on the longest focal length to take.

Do not despair if no telephoto lens gives sharp pictures. Some cameras are quite incapable of giving sharp images because of camera shake. That is the time to consider the other end of the focal range. Many of the more interesting or artistic pictures are taken with wide-angle lenses. These take in the true scenic atmosphere, and can be quite useful scientifically. The Moon's shadow will show and some of the fascinating colours around the horizon. Many of the best pictures are taken on a 28-mm lens, but a 21 is becoming quite popular. Fish-eyes are expensive, and are rarely used except for these special events and meteor work.

Do not forget the standard lens fitted to the camera. This is often the best choice for a very simple reason – it is, by its very nature, designed to give the best all-round results! Always take the standard lens, even if you have no intention of using it. If all the fancy lenses fail, get damaged or stolen on the way, you will get something at the end. Just keep the standard lens on the camera until you get there and do not let it out of your sight.

Hiring equipment is always fraught with dangers. For the majority of people it is not recommended unless you are able to afford to hire long enough for trials. Hiring is best for the professional already familiar with a camera system.

Choice of Film

There are so many good films on the market now that it does not matter what you use. If you have a favourite for normal use, stick with it. The film's properties will be known and it is much safer.

The general rules are very simple. The slower films give better grain and colours. The faster films are necessary to record faint detail where quality is of secondary importance.

Monochrome

Black and white films are of limited use except on tight budgets and specialized scientific work. Do not use them unless you have those special reasons.

Print

Print films are the best choice for those needing to show pictures to friends and nothing much else. A major snag is in getting decent prints because of the unusual background. If you have a good printer, or can do the job yourself, print film is preferred to slides because of its versatility.

Slide

Colour reversal film has instant appeal, and a large projected image does bring back much of the eclipse atmosphere. Reversal film is also preferred for magazine and book reproduction. Consequently, many of the more serious photographers use only slide film. The choice then falls to personal preferences.

Kodachromes have a major advantage in a tried technology

which has been around for over half a century now. Just about every other film seems to change faster than it can be reviewed. As with the equipment, check out a particular film and if you like it, buy a bulk of a particular batch. This is vitally important as even batches of the same film can have differing effective speeds and colour rendition. Keep the film in cold storage until needed.

How much?

How much? is not as critical as 'what?'. The biggest mistake to make is to take too many types of film. Even with film reminders fitted to the camera, it is all too easy to load the wrong type of film. Worse still, to forget what is already in, and have to change in a hurry or lose valuable pictures whilst doing so.

The best advice is only to use two speeds of film. One for everything, the other for emergencies or for special experiments. If you are in the habit of using a mix of slide and print film, the same rule applies. Then it will not matter what type is loaded, the exposures will be the same. The more cameras you have, the more critical this rule becomes.

100 ISO plus some 400 is the most popular choice. Film of these speeds can usually be bought in local shops.

How much is quite easy to answer on eclipse trips. It is – at least twice as much as you think you will need, *plus* one reel for the eclipse specially put aside and labelled.

Storage

At one time this was not important. Modern films are so fast and airport security now so strict that great care must be taken. Set aside all the film you think necessary for the eclipse itself and store separately. Many airports will not allow hand searches and this is another good reason for using only slow film for everything else as it is much more tolerant of X-ray examination.

Owners of 120 film cameras have no worries. Simply take the reels out of the metal foil and carry in pockets. These will not show on metal detectors.

Single reels of 35-mm film can be carried in the same way, perhaps up to three. This dodge will work provided all keys and other loose metal objects are kept in baggage. The really keen photographers, like myself, will use only re-wound film in plastics cassettes. That way, a large number can be carried on the person through a body search and not upset metal detectors.

Count Down for Photography

As has been said in 'Preparations for Photography', practice is all important. *Everything* must be practised and practised long before leaving home.

Particular points to watch are that you do not become too ambitious and try to run too many cameras. Aim to get one good picture at 2nd and 3rd contacts for Baily's Beads or Diamond rings, and just two during totality. One of these should be a short exposure for prominences, the other a long one for the corona. Anything else is a bonus, *not* the main objective.

Elevation

The first thing to do before exposing any film is to find out the anticipated elevation of the Sun at total eclipse. Most problems arise from not checking this vital piece of information.

The next two eclipses are at the extremes. Finland in 1990 has a partially eclipsed sunrise with totality a degree or two above the horizon. Mexico in 1991 has a maximum elevation of 89.7°! Other sites on the tracks must be checked over.

Test Assemblies

Then assemble the complete set-up indoors with an imaginary N/S line. Elevate to the anticipated angles and squint through the viewfinder. In the comfort of your lounge is the time and place to discover that the new lens snags the tripod, or that it is impossible to get a decent view without a crick in the neck. Think about the comfort in a wet field or on blazing sand, or wherever, and re-design the set-up if necessary. Investment in a right-angled viewfinder may be all that is necessary.

The next stage is to practise and time the complete assembly from the kit bag – in total darkness. Many eclipses take place just after dawn, and the 1990 one in Finland is a classic case. Use no more than a hand torch. If you have an assistant, so much the better. Your practice runs will show up assembly problems and what pieces are likely to get dropped. Again, a carpet is much better than mud and sand.

Viewfinder

A wide-angle lens obviously has no problem in lining up. The same is not true of long telephotos and telescopes. Lining up on the Sun with a heavy filter is surprisingly difficult, particularly in the haste often necessary on a crowded eclipse site. Time and effort designing some form of finder is well spent. It could be a back-up camera which is used as a counterweight at the same time. Check that you can align them accurately and that they do not snag vital parts at the anticipated elevation. Pay particular attention to balance and drive irregularities.

Test Stability

We still have not got to the stage of film exposure.

Whilst still in your lounge, with the set-up at the correct elevation, focus as best you can on whatever is there (blurred because of its closeness). Tap the lens. Does the image take longer than 1/10 second to settle down? If so, the structure is unsuitably flimsy. Try the effect of adding weights or bracing struts. Also try the effect of a hairdryer at close range to simulate strong wind buffeting.

The taller the tripod, the worse it is. Similarly, the longer the focal lengths used, the more troublesome tripod 'flop' becomes. A custom-built, short wooden job is often best.

There is an old rule in photography with SLR cameras governing hand-held pictures. What this means in practice with long focal length lenses is that exposures in the most useful range of 1/125 to 1 second give most bother. Try this range (camera empty but with a dummy film in to get the balance right). If the image jumps about after exposure, camera shake is a real problem.

Check the mounting for slackness and any imbalance and correct if necessary. If this does not improve the situation then you will never get a really sharp image with that camera/lens/tripod combination and test exposures are a waste of time.

Test Objects

Whilst test exposures of the Sun are important for the partial phases, they are not a very good final check on tripod and camera stability. Exposure times are too short even with a decent filter.

By far the best test object is the Moon. Not only is it the right size, but it is also the right lower range of brightness. Although its apparent motion is slightly less than the Sun, it is close enough to check guiding or motor-drive errors.

Full Moon is about the brightness of the inner corona. A few exposures will be a useful double check that the f ratio of the optical system is what is claimed. Full Moon, is, by definition, a day(Sun)lit object. Correct exposure at f 16 is given by: 1/the film ISO(ASA).

149

The crescent Moon is the best test object for focus and camera shake checks. Allow two stops extra exposure to pick up crater detail, and do duplicate exposures during periods of good seeing. It will probably take some lunations to iron out all the bugs.

Do not always expect that the image in the viewfinder is absolutely correct focus. Make test shots a little out of focus either way and see if it improves the image sharpness. If it does, make a note on the lens where *actual* focus occurs. Does this change each time you re-assemble the set-up?

Exposure Planning

Work out a planned sequence of exposures. Keep it simple and write it down. Work through this plan in a dummy run (no film in the camera), with no more light than the full Moon. Allow at least 50 per cent of the time to just stand and stare. Use that 50 per cent now to note the time taken and write it against each step. An autowind camera with an intervalometer does make life easier these days.

Keep at it until you have a plan capable of being carried out in the time of totality. Simple plans can be committed to memory, but the panic of last minute hitches makes this a dangerous policy. It is much better to take along a simple 'Walkman'-type player. Use your script and a stop watch and actually record the plan, in real time, and work through the sequence using your own voice prompts. If you are well away from others on the site, another recorder to take down your verbal records of what you *actually* do on the day is a great idea. However, do bear in mind that other workers might not like your commentary. The advantage of your taped prompt is that it is directly into the ear and blots out other voices.

Suggested Plan

1. Where time allows, always take plenty of pictures of the site as you arrive and set up. Most eclipse viewers pack up immediately after totality and are never seen again!

2. If you only have one camera, change film at the first opportunity to the right one, check the settings, then leave well alone.

3. Set up the equipment. Check that moving bits work and rectify problems. Use your audio prompt.

4. If you want pictures of the partial phases, take the absolute minimum needed to check that the camera is working properly. Reserve the partial phases to after totality.

5. One minute before totality, double check that the camera is correctly set for Baily's Beads.

6. Half a minute before totality, remove the solar filters, lens caps and other shields.
This is an absolutely vital step in your script.
The camera will come to no harm now, but do not look through the viewfinder yet.

7. Still not looking through the viewfinder, judge when the final traces of sunlight twinkle out, the Baily's Bead stage, and take that picture.

8. Wind on without looking at the camera but the eclipse itself. The audicue should be in synchrony. Many prefer to stop the tape early and restart now. Local timings can be out by several seconds, particularly if the site is moved.

9. Look around and at the totality, using optical aid if taken.

10. Take another picture with the same settings as Baily's Beads. This will record the prominences and inner corona. Re-wind and re-set the camera immediately for a long exposure.

11. Re-look at the eclipse, and general scenery.

12. About half-way through, take the long exposure and repeat as time allows.

13. Re-set for the Baily's Beads exposure.

14. Re-look about and wait for the Diamond Ring effect.
Do not attempt to look at this through optical aid or the camera viewfinder, as **it is dangerous**.

15. Savour the gorgeous sight with the naked eye, and trip the camera shutter immediately it is seen.

16. Wind on as rapidly as possible and take as many pictures as you can in about 5–10 seconds.
The camera will come to no harm.

17. Relax.

18. Re-set the camera for partial phases and finish the film on these.

19. Double check dismantling and packing, audicueing if necessary.

20. Double check again that films are correctly labelled and stowed away for the journey home.

This suggested plan is amongst the simplest, but, even then, has twenty important steps to carry out. The need for practice beforehand should be abundantly obvious by now!

The inexperienced viewer will spot a major problem in this plan – How do you know when eclipse totality starts and ends, the 2nd and 3rd contacts?

Most expeditions normally have a seasoned viewer or astronomer with them with accurate timings and means of checking. Make sure that some plan is agreed within the party for timing warnings to be given out. This must be done several days in advance, preferably before leaving home. Your audio cue can then be synchronized from this. A one- or two-minute warning is usually enough.

Should you be on your own, or wish to do it yourself, a good way of preparing is to use a normal room light fitted with a dimmer. Get someone to turn the control from full to nothing in about twenty seconds and this will give a fair approximation to how the final light drop appears in fact. The final 'winking-out' can take up to ten seconds depending on the lunar limb shape, and information on that is also often available on site. Cameras with autowind can be set going during this period, but stop as soon as totality starts to conserve film. The end of totality is seen in two ways, assuming no timings. The easiest is to keep a close view of the corona on the opposite side from the last vestiges of sunlight. Binoculars will show a perceptible lightening of the corona as totality ends. As soon as a perceptible 'pinkening' occurs, totality is almost over and optical aid should be stopped immediately. **This is a dangerous procedure**.

The safer and more spectacular way is to keep an eye on the horizon from which the Moon's shadow came. The brightening of the skyline there is quite remarkable. It arrives at an amazing rate, but do not look too long or you will miss the Diamond Ring effect as the first sliver of sunlight reappears.

EFTS, or the 60-16-60-6

Knowing what to photograph at an eclipse is always the most difficult thing to decide. Quite apart from the weather uncertainties, there are just simply too many variables of equipment to be dogmatic. If you can lay your hands on one of the many published lists of exposures for the different phenomena, the recommendations are usually spot on.

What this boils down to in practice is that virtually any exposure will give a result of some sort, because of the wide latitude in modern films. It is also important to remember that the inner corona is about the same intensity as full Moon, which is, in turn about the same intensity as an average daylit scene. Excellent pictures of the inner corona and fainter prominences are taken by setting the camera for a daylit scene and leaving the camera controls alone thereafter.

This makes calculation easier and the following mnemonic is suggested as a simple way of doing last minute adjustments:

E for Exposure Index (ASA/ISO)	60
F for focal ratio	16
T for time of exposure	60
S for stops of brightness range	6

E = The nearest film speed to 60 is 64. This fits Kodachrome 64 (or Fuji 50D).

F = The focal ratio of many long telephoto lenses is around f16.

T = An exposure of 1/60 second for the inner corona is right in the middle of the average camera shutter speed range. An error of a setting either way is not too serious, because:

S = This is the normal brightness range in stops for the whole visible eclipse. It is also a similar range to the response capability of modern print films.

The brightness range is from the bright prominences to sky background where the outer corona peters out. If we set the inner corona brightness slap-bang in the middle, which it is, this gives a tolerance of 3 stops either side, or the 6 stops in all.

Practical examples will show how it all fits together, remembering that approximations are more than acceptable.

Example 1

We have only 100 ISO film, our mirror lens is f8, and we want as much corona as we can get. Using quick mental arithmetic we have:

E = +1 (100 is closer to 120 than to 60)
F = +2 (f8 is 2 stops faster than f16)
T = The figure we need
S = −3 (3 stops fainter)

The new exposure time is +3 stops −3 stops, or no change. That means that an exposure time, T = 1/60 second will record much of the outer corona, burning in the prominences.

Example 2

Put another way, what film should we use with the same lens, using 1/1000 sec to minimize camera shake on a poor tripod?

$$E = \text{The figure we need}$$
$$F = +2$$
$$T = -4 \ (1/1000 \text{ is } 1/60 \div 2 \div 2 \div 2 \div 2 = 4 \text{ stops})$$
$$S = -3$$

This means we need a film 5 stops faster than EI (ISO) 60 or around a speed of 2400. There are now films about faster than this at 3200 ISO, and they will do nicely.

Once this relationship has been established, it is worth considering a slightly longer exposure time of 1/500 sec with the better quality of the many 1000 ISO films about.

Example 3

We have Kodachrome 25, our telescope is f16 and we want good prominences and Baily's Beads.

$$E = -1 \ (25 \text{ is close enough to } 30)$$
$$F = 0 \ (\text{no problems here})$$
$$T = \text{The figure we need}$$
$$S = +3 \ (3 \text{ stops brighter})$$

This gives a +2 stops speed correction, or $1/60 \div 2 \div 2 = 1/250$ sec exposure time for the prominences.

For the inner corona, exposure time is $1/250 \times 2 \times 2 \times 2 = 1/30$.

For the outer corona, exposure is $1/30 \times 2 \times 2 \times 2 = 1/4$ sec.

It is important to calculate the range of exposures likely over the whole eclipse phenomenon, because there is a final factor to take into account in the form of cloud cover. Again, a range of 6 stops is possible. Cirrus-type cloud cover is quite common and 1 or 2 stops will take care of that. In Example 2 for instance, the extra 2 stop light loss due to cirrus means that much of the outer corona will be missed. There will also be a severe image quality degradation. However, it is worth a try, and that means that instead of 1/4 second exposures we need 1 second, about the limit of many cameras on their set range. Where even longer exposures are needed, a cable release and personal timing may be necessary. Do not attempt accurate timing by looking at a stop watch. The old meteor watchers and photographers' rule of 'one thousand', 'two thousand', and so on to count off seconds is quite good enough. You simply cannot afford to waste valuable time staring at a watch when Nature's Treat is on view. Forget the whole idea of long exposures, use an audio prompt, or opt for shots of the inner corona or prominences only. If the Sun can be seen in the few seconds before totality, you will get something.

The Golden Rules of Eclipse Chasing

The Golden Rules of eclipse chasing boil down to five DON'Ts:

1. DON'T be too ambitious.
Take photographic equipment, but be prepared to get nothing from it and just use the 'Mark 1' eyeball, preferably with binoculars.

2. DON'T change an established system.
If you have tried and trusted equipment, stick with it. The eclipse site is not the time and place to find the limitations of the latest 'Gizmo' with flashing lights.
'System' also means procedures. To err, is human . . .

3. DON'T panic.
No explanation needed. See Rule 1!

4. DON'T lend anything to anyone.
Lend equipment after the eclipse by all means. Before that, Jack really is Number One.

5. DON'T let your film and pictures out of your custody.
Your pictures are a unique record and memento. It is as near to a certainty as anything can be in this world that if you lend original pictures, they will either be lost or be returned damaged. Make copies specially for loan or reproduction.

Check List of Items to Take on Eclipses

This check list is laid out in a suggested order of priorities, based on experience. Unless you are really dedicated, do not work through the main list beyond Item 6, until you pick it up again at Item 32. The dedicated should make a copy of the full list, use it as a work programme, and select those items which fall into their own photographic plans and order of priorities. Add items which are important to your programme.

Item	Check here	Item	Check here
1.	Binoculars and/or Telescope	18.	Battery operated tape player and headphones
2.	Solar filters for naked-eye viewing of the partial phases		Detailed written observation check list regardless
3.	Celebration necessities (see Item 35)	19.	Battery operated tape recorder
4.	Camera bodies		Note book and pencil regardless
	Still		Preferably, both on a lanyard round neck
	Ciné	20.	Spare tapes for recorders
	Video	21.	Spare batteries for recorders
5.	Standard lens for each type of camera	22.	Small torch, with lanyard to hang round neck
6.	Film for each camera	23.	Spare batteries for torch
	100 (preferably slower) ISO	24.	Puffer brush for cleaning lenses
	400 ISO	25.	Lens cleaning tissues or cloth
	Specials	26.	'Cotton Buds'
	Video tapes	27.	Reel of adhesive tape
7.	Special lenses for each type of camera		Clear adhesive
	Telephoto		Masking
	Wide-angle	28.	'Blu-Tack'
8.	Special mounts for lenses to cameras	29.	Tweezers
9.	Lens cap for each type of lens	30.	Jeweller's screwdriver set
10.	Lens hood for each type of lens	31.	Set of spanners for tripod and other gadgets taken
11.	Solar filter for each lens	32.	Sunglasses
	Is it firmly attached in a full gale?	33.	Sunhat
	Can it be removed and replaced quickly?		Proper clothing for eclipse site, including shoes
	White cloth to put over equipment	34.	Seating
12.	Spare batteries for each camera body	35.	Refreshments
13.	Spare batteries for each motor drive		Fresh fruit juices are best (until after third contact!)
14.	Sturdy tripod for each type of camera		**For the Rest of the Trip**
15.	Spare batteries for tripod equatorial drives	36.	Medical necessities
16.	Cable release for each camera	37.	More film
17.	Instructions (copies for each and EVERY piece of equipment)	38.	More tapes
		39.	Flash gun
		40.	And yet more batteries!

Pay particular attention to battery power. So many items in this list take batteries that a considerable proportion of the weight allowance will be taken up with them. Consider very carefully the advantages of primary batteries which can be bought cheap and left behind, against re-chargeables which can be 'topped up' from the local mains just before the eclipse.

Answer to query on page 135.
The first year of the graph is 1960. Thus readers might have chosen the peak years of 1965–1966 by mistake if they were looking for an increase in the accident rate. The years 1974–1975 might have been chosen by those looking for a decrease. Consider the numbers killed. From 1966 to 1968 the annual number fell by over 1,000. From 1968 to 1969 there was a sharp rise, which then continued at a smaller rate up to 1972, whereafter there was a decline. The figures for those injured exhibit a similar pattern apart from the deviation between 1970 and 1971. Obviously there is no justification what-soever for the claim of a lower accident rate in 1969 and 1970. The reader must judge for himself how strong the evidence is in favour of the author's contention that the introduction of C.E.T. is likely to increase the number of casualties, even if only marginally.

The Great Attractor

PAUL MURDIN

Is the Universe the same in all directions?

The Universe is mostly empty space; galaxies mark the distances in the Universe like surveyors' stakes on a building site, not like bricks in a cosmic house. So it is not surprising that theoretical cosmologists (who set themselves to form theories of the Universe as a whole) look at the Universe in a geometrical fashion, as a surveyor looks at a site plan. Cosmologists talk about space and time and curvature – these are geometers' terms.

Indeed, books on theoretical cosmology tend to be written in the manner of Euclid's book on geometry, with axioms and definitions from which follow theorems and special cases. Of course, every astronomer realizes that the theoretical Universe is at best only an approximation to the real one and that the assumptions must be true only to a limited extent. In the same way, the surveyor may use Pythagoras' theorem to calculate the third side of a right-angled triangle on a plan, knowing that, for various reasons (for example, we live on a spherical Earth, not a flat one), the calculation will not give exactly the same answer as if he measured the side.

Astronomers are just like all scientists when they check the cosmologists' calculations against the real Universe. Observation is the test of whether the theory is an accurate description. The trouble is that cosmology is about the whole Universe, which is big and contains a lot to study, most of it very far away and faint; so observational cosmology is a difficult subject.

One of the assumptions which is built in to theories of cosmology is called 'isotropy'. It says that the Universe is, by and large, the same in every direction. Everyone has realized that this can only be true if you ignore local details. For instance, on a scale of metres nearby to where you live, there is obviously an 'up' direction and a 'down' direction, and they are different; kick out from under you the chair on which you sit to read this article and you will automatically select the 'down' direction in which to move, not the 'up'! On a scale of millions of kilometres, there is similarly a difference between the direction towards the bright Sun and a direction out to the far dark regions of the Solar System. On a scale of thousands of light years, there is no isotropy in our Galaxy, which has a centre and a rotation plane. But it had been assumed that on a scale of millions of light years the Universe was isotropic, with one direction as good as any other.

This was the simple view. Of course, anything initially looks simple when viewed from megaparsecs away (1 megaparsec is a million parsecs, about 3 million light years). The simple view was an initial working assumption; it got astronomers started on the subject, and, later, everything turned out to be more complicated.

The Hubble flow

Astronomers have recently been finding out that the real Universe is measurably anisotropic: there is an 'up' and a 'down' to the Universe, when measured on a scale of hundreds of millions of light years. The cause of the anisotropy seems to be an object called the Great Attractor, and it is pulling us, and our Galaxy, and all the galaxies in our neighbourhood. It is from its effect on all the nearby galaxies that it makes its presence felt.

Sixty years ago, the American astronomer Edwin Hubble discovered that the Universe was expanding. He measured the distances and redshifts of galaxies and found that the more distant a galaxy was, the faster it was receding from us. He expressed this as a simple formula known as Hubble's Law, $V = Hr$, where V is the speed of a galaxy away from us (its redshift) and r is its distance; H is a constant known as 'the Hubble constant' and it has a value of about 100 km/sec per megaparsec (some authorities prefer a lower value). This means that a galaxy which is, say, 10 megaparsecs away has a recession speed of 1,000 km/sec and one which is twice as far, at 20 megaparsecs, has twice the speed, or 2,000 km/sec. Cosmologists expect that Hubble's law breaks down at very large distances from us (it doesn't apply to quasars, for instance), but it is a good approximation for the galaxies which we can actually see.

The Big Bang

Hubble interpreted his law as the consequences of a big explosion, the Big Bang. In a mass marathon race, started at the gun, there is turmoil at first, but the faster runners draw away and move to the front: the faster runners quickly become the most distant. So it was that Hubble visualized the expansion of the Universe.

The flow of the galaxies in the expansion of the Universe is quite accurately isotropic. In other words, the Hubble constant, H, is very similar in every direction. If you plot the distance of galaxies against their redshifts for a sample of data gathered by the Anglo-Australian Telescope for galaxies in the southern sky, you find a value for the Hubble constant which is not much different from the value for a sample of galaxies observed in the northern sky by the telescopes in La Palma. But the definiteness about this statement has, up to now, been limited by the number of galaxies which have been studied and the accuracy with which they can be studied – if there are few galaxies in each sample it is not surprising that you cannot spot directions in which the Hubble constant is 90 km/sec per megaparsec and others where it is 110 km/sec per megaparsec.

The Seven Samurai

In the last five years it has been possible to study the flow of the galaxies in the explosion of the Big Bang in a detailed way, as if investigating the course over which a marathon is run, seeing where the individual racers run faster on the down-hill sections and slower on the up-grades. The detailed work investigating the flow of galaxies in the expansion of the Universe has been carried out by a group of astronomers who became known as the Seven Samurai, after the Japanese warriors in the film by Kurosawa. They were David Burnstein (Arizona State University), Roger Davies (formerly of Kitt Peak Observatory and now at Oxford University), Alan Dressler (Mt Wilson and Las Campanas Observatories), Sandra Faber (Lick Observatory), Donald Lynden-Bell (Cambridge University), Roberto Terlevich (Royal Greenwich Observatory), and Gary Wegner (Dartmouth College). (How they got together is, like the first half of Kurosawa's film, a long story!) They travelled to virtually every large telescope in the world, determining fundamental data on hundreds of galaxies.

A distance indicator

The Seven Samurai chose to study elliptical galaxies, the commonest, and, in many astronomers' eyes, the most boring galaxies in the Universe. The Seven Samurai reasoned that they wanted to use galaxies which really represented surveyors' stakes peppering the building site of the Universe, not flagpoles which were widely spaced and which, although they may have carried attractive flags, could all be special in some distinctive, individual way.

Figure 1. Elliptical galaxies (like NGC 1399 – the larger – and NGC 1404) are fuzzy, spherical or elliptical collections of stars. Notice the globular clusters which swarm about NGC 1399 in this Anglo-Australian Telescope photograph by David Malin.

The Seven Samurai discovered a way to mass-produce measurements of the distances of the elliptical galaxies. They realized that if they could estimate the diameter of an elliptical galaxy (in, say, kiloparsecs) and at the same time measure its angular size in arc seconds, they could then accurately estimate its distance: the further away a galaxy of given size is, the smaller it appears. Measuring the galaxies' angular sizes was relatively straightforward: they could take pictures quickly with CCDs. (CCDs are Charge Coupled Devices, small arrays of diodes which convert light to electric charge and which can be read out very accurately to give a television-type picture.)

Measuring the elliptical galaxies' real size was more difficult, but the Seven Samurai discovered that they could do it from the same data with which they measured each galaxy's redshift, namely its spectrum. The spectrum revealed not only the speed with which the galaxy of stars was moving in bulk – its redshift – but also the speed with which the stars were moving within each galaxy. It was already known that there was a straight-forward relationship between the brightness of elliptical galaxies and the speed with which stars moved in them (the Faber-Jackson relationship). Roughly speaking, the Faber-Jackson relationship measured the brightness and the total mass of all the stars in each galaxy. The Seven Samurai found that there was a more complicated relationship between the speed of the stars in an elliptical galaxy and its diameter measured in a special way. Although it was a more complicated relationship, the new formula served to determine the distance of a given elliptical galaxy more accurately (plus or minus 15 per cent). When several elliptical galaxies formed a cluster, the distance to the cluster could be found to an accuracy of 3 per cent. This gave the accuracy needed.

Streaming motions

Altogether, the Seven Samurai took spectra and CCD images of about 500 elliptical galaxies. They used the equivalent of half a year on virtually all the large optical telescopes in the world; in financial terms it was a multi-million pound project. They observed elliptical galaxies in all parts of the sky save those regions hidden behind the dust of the Milky Way. They processed the data to determine the distances and redshifts of all the galaxies. They then determined the average Hubble constant for all the galaxies together in the sample, to measure the Hubble flow as it appeared on average all over the sky. Finally, they plotted each galaxy in space, and looked at how the speed of each departed from the Hubble flow in its neighbourhood. They were looking in effect at the flow of individual streams within the motion of a bigger sea.

Figure 2. Elliptical galaxies dominate this picture of a rich cluster of galaxies CA 0340-538. Most of the images in the picture (except the small hard circular images of stars) are galaxies – the cluster had hundreds of members.

The Seven Samurai noticed that galaxies in the constellations Perseus, Pisces, Fornax, Pavo and Indus (in one half of the sky) were, in general, receding at the Hubble flow. In Virgo, Hydra and Centaurus, in the other half of the sky, the galaxies were receding somewhat more than the Hubble flow, and so streaming away from us.

The Seven Samurai interpreted this to mean that all the galaxies nearby to us – in effect, all the galaxies in a huge volume of space 200 million light years around us – are streaming in bulk in the direction of Centaurus, near to the direction of the Southern Cross. Effectively this is the downhill direction in the flow of galaxies in the nearby Universe. We ourselves are falling 'downhill' with a speed of 700 km/sec. The Centaurus cluster is falling 'downhill' even faster. The Perseus cluster is not falling towards Centaurus very fast at all.

The Great Attractor

Why should this be?

The simplest explanation is that somewhere in the 'downhill' direction, at a distance just beyond the region of space which the Seven Samurai surveyed, is a large mass concentration, which one of the Seven Samurai, Alan Dressler, termed the Great Attractor. It gives galaxies near to us an extra pull of gravity in that direction. It has a mass of somewhat over 10^{16} solar masses (this is the mass of 10 thousand million million suns) and it lies at a distance of some 500 million light years from us, beyond the Centaurus cluster of galaxies. A galaxy contains 10^{11} solar masses; the previously known biggest things in the Universe were rich clusters of galaxies like the cluster in Virgo and they contain up to 1,000 galaxies (10^{14} solar masses).

The Great Attractor is 100 times bigger than this – it is the biggest thing in the Universe.

The cosmic microwave background

There is independent evidence for the remarkable discovery of the streaming motion, which might be caused by the Great Attractor. It comes from observations from the cosmic microwave background. This is the name for the present appearance of the cosmic fireball of the Big Bang.

The cosmic fireball of the Big Bang was generated in the early history of the Universe. Astronomers see its remains as a faint whisper of microwave radiation, discovered in 1964, bathing the

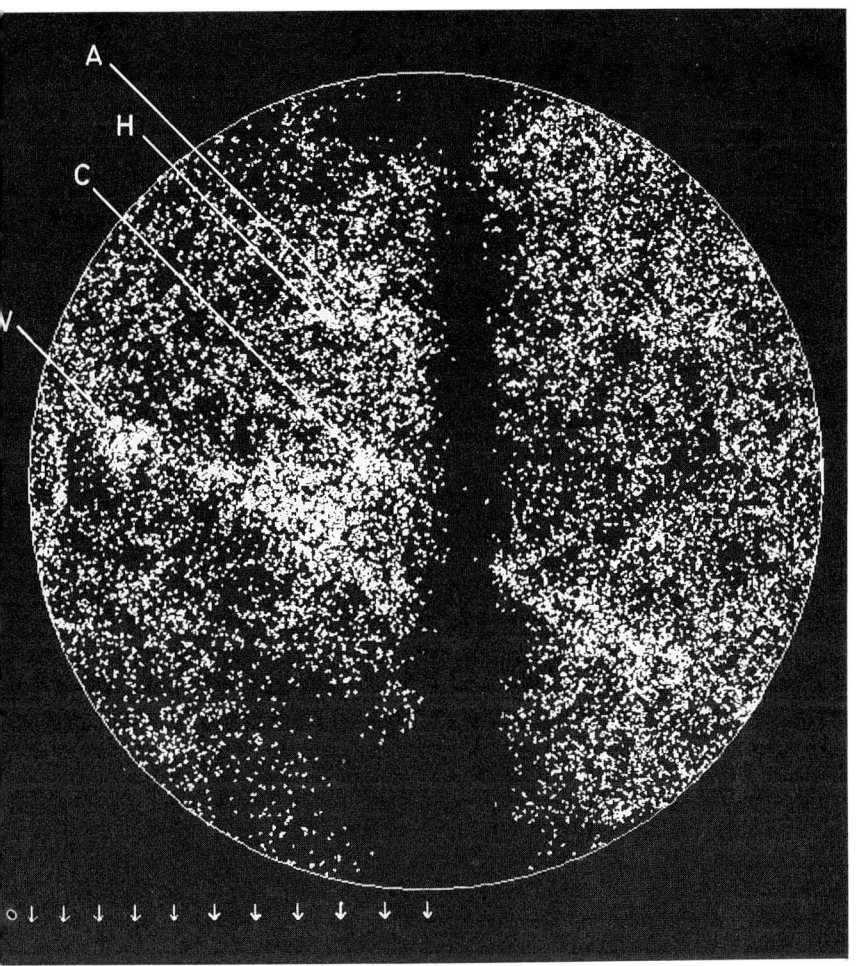

Figure 3. Computer plot of the supergalaxy made by Lahav from the ESO catalogue of galaxies, supplemented by the UGC for the north and the Morphological Catalogue for the $-2° > \delta > -17°.5$. The picture is centered on the direction of the streaming, $l = 307$, $b = 9$, and contains a hemisphere of the sky in an equal area projection. The zone of avoidance caused by the Galaxy is the dark vertical band. The Galactic center lies 37° up from the bottom. The Virgo, Centaurus, Hydra, and Antlia clusters are indicated. The great concentration of galaxies just below the Centaurus cluster is obvious. The nonlinear scale indicates 9° radial intervals; the arrows at the left are appropriate to the edge of the picture, those at the right are appropriate for the center.

Earth from all directions, like a dish heating in a microwave oven. The expansion of the Universe has by now cooled the radiation of the fireball down to only 2.756 degrees above absolute zero.

The cosmic microwave radiation is thus a fossil of the Big Bang. With one exceptional feature, it is remarkably uniform. The most accurate recent measurements were reported in 1988 by Rod Davies of Jodrell Bank, using a special radio telescope set up in the shadow of Teide in Tenerife, in the observatory of the Astrophysics Institute of the Canaries.

The cosmic microwave radiation is slightly hotter in one direction than another. Its temperature seems to be 0.0033 degrees more in a direction towards Hydra. This can be interpreted as the direction towards which our Galaxy is moving at a speed of some 600 km/sec. This is not quite the same direction nor the same speed as expected from the pull of the Great Attractor.

However, the Great Attractor is not the only object which pulls our Galaxy. The nearest big cluster of galaxies to us is the Virgo cluster. Part of the motion of our Galaxy is calculated to be due to the tug of the Virgo cluster, although it is not obvious from the Seven Samurai's data, collected over a much larger volume of the Universe. When this pull is corrected for, the streaming motion representing the pull of the Great Attractor and the motion of our Galaxy through the cosmic fireball agree.

The nature of the Great Attractor?

Unfortunately, the 'downhill' direction of the streaming motion lies in the Milky Way, in one of the dustiest regions (the Southern Cross is well known as the constellation which contains the Coal Sack, an especially black patch of dust). So the nature of the Great Attractor is somewhat obscured. It is probably not a single object, like a frighteningly massive black hole. It seems to be an especially rich concentration of galaxies. Counts of faint galaxies in this direction do seem to show an excessive number, although the result is indeed somewhat confused by the Milky Way. The Great Attractor may be a super-super-cluster of galaxies, perhaps.

If it really is as massive as it seems, then just as the Great Attractor attracts galaxies in our region of space, it must attract galaxies all round itself. Galaxies beyond 500 million light years in the direction of Centaurus must be falling back towards the Great Attractor. Astronomers want to test whether this is true by looking to see if distant galaxies in Centaurus are streaming towards us, at twice the speed that we are streaming towards the nearer galaxies in that direction.

It is difficult to make these observations because the galaxies are so faint. However, galaxies in distant clusters are concentrated into relatively small regions of the sky, and astronomers have been working on ways to observe large numbers of them at the same time. They use optical fibres to pick up the light from scores of galaxies within the field of view of a big telescope and feed the light from all the galaxies at once into an instrument to record their spectra. The individual spectra of the faint galaxies take a long time to record, and it is time consuming to construct the fibre feeds and position the telescope accurately to pick up all the galaxies on all the fibres. None the less, so many galaxies are observed at once that the extra time is well spent.

Working at the forefront of technology, using new instruments like this, and motivated by the desire to seek fresh knowledge about the Universe in which we live, astronomers in 1989 have been trying to confirm whether the Great Attractor really exists. Perhaps the results will be becoming known by the time this *1990 Yearbook of Astronomy* is published.

More Galaxies than Meet the Eye

MICHAEL DISNEY, STEVEN PHILLIPPS
and JONATHAN DAVIES

When nineteenth century astronomers managed to determine the distances to the stars they found that the bright naked-eye stars, such as Rigel and Betelgeux in the constellation of Orion, were generally intrinsically much more luminous – perhaps thousands of times more luminous – than our Sun. Indeed, none of the stars brighter than apparent magnitude 2 is actually dimmer than the Sun.

Does this mean, then, that the Sun is a highly insignificant star and that 'normal' stars are very much more luminous objects? In fact, the answer is no, and the problem is an instance of what is called 'observational selection'. To see this, consider as an example a set of stars all at the same distance from us. If there is just one bright star and lots of intrinsically fainter ones, it is none the less the one bright one which we will see as the naked-eye object, while the others are not seen at all. If we only look at the objects which are easiest to see, then we may get a highly biased view of what is really out there, and this has certainly turned out to be the case for stars, as it is now known that the majority of stars are in fact much less luminous than the Sun. Very powerful stars such as Rigel are really extremely rare, but they force themselves on our attention by being easily visible despite being quite distant. On the other hand, a dim red dwarf millions of times fainter can only be seen if it is in our own cosmic backyard, so to speak.

What, then, of the population of galaxies, huge systems of stars outside our own Milky Way? Can the same thing be happening there? Might the well-known giant galaxies like M31 in Andromeda or the 'Whirlpool Galaxy' M51 actually be the exception rather than the rule, with the Universe really being full of large numbers of less conspicuous systems? What does observational selection do in the case of galaxies? In fact, observational selection turns out to be even more insidious for galaxies than it was for stars, as we shall see below.

First, exactly as for stars, intrinsically faint galaxies can only be seen if they are quite nearby (by intergalactic standards). The Magellanic Clouds, for example, which are our nearest galactic neighbours and are easily visible to the naked eye from the Southern Hemisphere, are quite modest galaxies. The Large Magellanic Cloud, some ten times smaller than our own (Milky Way) Galaxy, and the Small Magellanic Cloud, ten times smaller still, are only so visible because of their proximity. If they were placed in the Virgo Cluster, the nearest huge aggregate of galaxies, they would be rather insignificant objects, and the many even smaller 'dwarf galaxies' known to exist in our Local Group of galaxies would be barely visible even with the largest telescopes. Clearly, then, we are biased against seeing low-luminosity systems, and we must make due allowance for this effect when determining the relative numbers of galaxies of different luminosities (the so called 'luminosity function').

However, in the case of galaxies, a further complicating effect comes into play. Because galaxies are spread out – unlike stars, which appear as tiny points – what really matters when detecting them is the amount of light they emit per unit area, that is their 'surface brightness'. Consider for example, a set of galaxies all of which have the same total luminosity (i.e. the same number of stars). If the stars are packed very closely together, the amount of light per unit area will be high, and so the galaxy will have high surface brightness. However, it will also be very compact, and unless it is very close to us it will appear very small in size and will be hard to distinguish from a star on, say, a photographic plate. (Stellar images have a measurable size too, typically two seconds of arc across, because of the blurring effect of the Earth's atmosphere; very bright stars' images are also smeared out by effects in the photographic emulsion.) On the other hand, if the stars are very spread out, the surface brightness will be very low, and the galaxy will be very difficult to detect against the general diffuse background of the night sky. Indeed, it is worth remarking that the night sky is not particularly dark by astronomical standards, even at specially selected observatory sites. In fact, since we are looking out of a 'lighted' spiral galaxy much like M31, it is not surprising that the surface brightness of our night sky is quite similar to that of 'normal' bright galaxies. Galaxies which have much lower surface brightness than the one we live in are going to be hard to see, because they will be drowned out by the sky glow even on the darkest night.

Clearly, then, both high and low surface-brightness galaxies will be discriminated against when we compile galaxy catalogues, because they are harder to recognize. In fact, we can quantify this ease of visibility as a function of surface brightness (essentially by working out how far away a galaxy can be and still be seen) and a calculation we have made is shown in Figure 1. Superimposed is the actual distribution of surface brightnesses among well-known galaxies. The correspondence is remarkable, and provides strong circumstantial evidence that what we normally observe are just the easy-to-see galaxies and that there may be many other, harder-to-detect galaxies which we know nothing about.

Hard evidence that such galaxies really do exist can come only from dedicated searches. Using the Automatic Plate Measuring

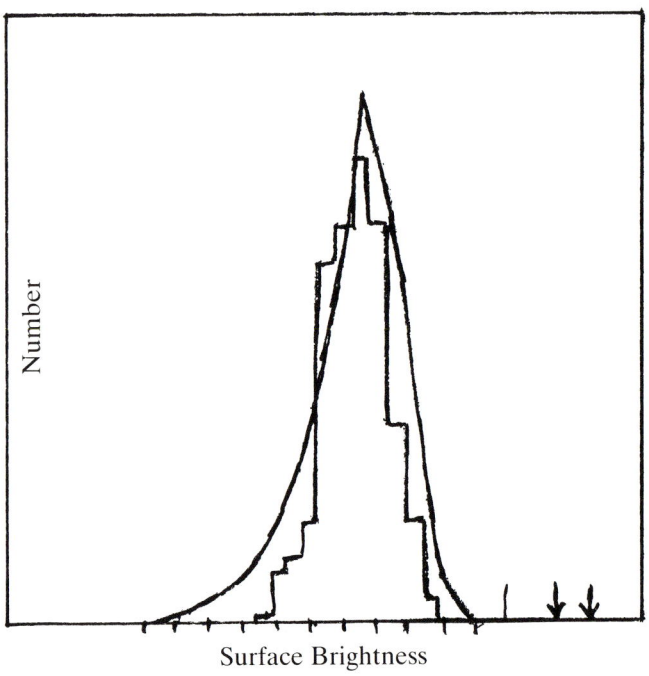

Figure 1. The ease of visibility of galaxies as a function of their surface brightness (peaked curve) compared to the actual numbers of well-known galaxies of different surface brightnesses (histogram). The two arrows show the positions of the lowest surface brightness galaxies known.

(APM) machine in Cambridge to measure plates of large areas of sky taken with the UK Schmidt Telescope in Australia has proved to be an excellent way to produce such surveys, since the computerized APM machine acts as a completely objective 'observer' and we can work out exactly what sort of images it will be able to detect. Indeed, a machine is indispensable for hunting for the very high surface brightness galaxies, since the problem is not in finding some objects but rather of picking out the few interesting ones from the vast numbers of almost identical images of stars. We are only just beginning this task, but already we have turned up a number of objects which were barely distinguishable from stars on a photographic plate, but which on detailed spectroscopic observations with the 3.8-m Anglo-Australian Telescope proved to be distant very active galaxies.

At the low surface brightness end, we can again make use of the machine to help us detect the very low contrast images we are looking for, since once the plate has been measured we can computer-enhance the pictures in order to show them up. In particular we can combine 'pixels' (i.e. picture elements) by a method called 'median filtering' which essentially finds the average brightness of a region while at the same time removing any small images (e.g. faint stars). This then shows up any patches of slightly higher brightness than the sky. We can then stretch the contrast until they show up clearly.

In fact, searching just one photographic plate of the Fornax Cluster area has turned up several hundred very low surface brightness galaxies, the most diffuse having surface brightnesses (at their centres) 100 times lower than galaxies like our own. David Malin, at the Anglo-Australian Observatory, has been able to see similarly faint galaxies by photographic contrast enhancement techniques, and one of these – named Malin 1 by his collaborators – has proved to be a really remarkable object. Although 100 times lower in surface brightness than say M31, it is over 100 times larger in area, so that the total amount of light it emits is actually greater. Indeed, it is one of the most luminous galaxies known.

To hunt for yet lower surface brightness objects, we need to move on to the highly efficient electronic detectors called CCDs (charge couple devices). Using such a device on the Isaac Newton Telescope at La Palma, we have been able to discover the galaxy with the lowest surface brightness disk yet known, about 250 times lower than the better-known galaxies.

Figure 2. The galaxy NGC 6822, one of the dwarf galaxies in the Local Group.

However, a disadvantage of CCDs is that they are very small; typically, fields of view are only 3 or 4 minutes of arc across, so that we can only survey the sky slowly in a piecemeal fashion. To cover the same area as our one photographic plate of the Fornax Cluster we would need around 10,000 separate frames, each taking maybe an hour of large-telescope time to reach the very faint limits we are interested in. Since this works out to be equal to every hour of 'dark time' (i.e. nights with negligible moonlight) for the next ten years, this is not a terribly feasible project!

Fortunately, though, there is a way of tackling this problem without stopping all other astronomers using the telescope, and this is the concept of 'parallel observing'. Most observations with a large telescope, such as the new 4.2-m William Herschel Telescope on La Palma, use only a tiny fraction of the available field of view, which is typically of order 1 degree across. For instance, if we are using a standard CCD we use only a few arc minutes of the field, as noted above, while if we are doing spectroscopy all we usually want is the light falling on the spectroscope slit, which may be around 2 arc seconds wide and 20 arc seconds long. If we can utilize some of the remaining surrounding area, then we can operate a secondary instrument the whole time that a scheduled observer is looking at an object near the field centre, without interfering with the primary observation. This is the principle of the parallel CCD camera, called the 'Hitchhiker' (after a suggestion by the Editor of this *Yearbook*!), which has several separate CCDs so placed as to collect the light from different parts of the periphery of the field of view. Since we will be using several CCDs at once and will indeed be able to observe every hour of every night, surveying the sky for exceedingly low surface brightness galaxies (or indeed any other hard to detect types of astronomical object) then becomes quite feasible. It is hoped that the 'Hitchhiker' will be in place on the WHT in about a year's time, so we should then begin to get an idea of just how many galaxies there really are.

Neptune from Voyager 2

PATRICK MOORE

Of all unmanned space-craft, Voyager 2 has without doubt been the most successful. Launched in August 1977, it by-passed Jupiter in 1979, Saturn in 1981, and Uranus in 1986, using the well-tried method of 'gravity assist'. Neptune, the outermost planet, was its final target. The encounter took place on August 25, 1989, and proved to be an unmitigated triumph. Though Voyager 2 was by then an old probe, it performed better than it had ever done before – an amazing feat in view of the fact that its main receiver had failed as long ago as 1978, so that it had been operating almost throughout its journey upon its back-up system.

Before the encounter, our knowledge of Neptune was decidedly limited, because no equipment on Earth will reveal anything but very vague markings upon its tiny blue disk. Arcs of rings had been suspected by the 'occultation method', and two satellites were known, one large (Triton) and one small (Nereid). Unlike Uranus, Neptune was known to have a source of internal heat, and was likely to be a more dynamic world, but that was about as much as we had ever been able to find out. The existence of a magnetic field was regarded as very probable, but not certain.

Even before the time of closest encounter, many fascinating discoveries had been made. Six new satellites (Figure 1) were found, all close-in and all small and dark – though it is true that one of them had an estimated diameter of 260 miles, so that it is larger than Nereid. However, it is too near Neptune to be observable from Earth, while Nereid has a very eccentric orbit and is sufficiently far away from Neptune's glare. (Nereid, in fact, was not well imaged by Voyager 2, because at the time of the encounter it was in the wrong part of its orbit.)

Next came revelations concerning the 'ring-arcs', which turned out to be the brightest segments of complete rings – though the entire ring system is very dim indeed, and most of it was only slightly above Voyager's threshold of visibility. Altogether there are three rings (Figure 2) and one broad sheet of particles. The arcs in the brightest ring contain objects a few miles across which have been called 'moonlets', though whether they are compacted bodies or mere clumps of particles remains uncertain. The Neptunian ring system is very dark; the particles making it up are no more reflective than coal. Two of them are associated with the newly discovered small inner satellites.

Radio emissions were detected as Voyager closed in, and the rotation period of the planet was finally fixed as 16 hours 3 minutes – rather shorter than had been expected. The magnetic field was a surprise. It is weaker than those of the other giant planets, but it is very appreciable, and the magnetic axis is inclined to the axis of rotation by 50 degrees, while the magnetic axis itself does not pass through Neptune's centre; it is offset by some 10,000 miles. A similar kind of inclination and offset had been found for the magnetic axis of Uranus, but of course Uranus' axis of rotation is inclined at more than a right angle (98 degrees), so that Voyager had approached its target pole-on, while the tilt of Neptune's rotation is only a few degrees more than that of the Earth.

Magnetically, then, Neptune is not unlike Uranus. It had been suggested that we had caught Uranus in the act of a 'magnetic reversal', but as Neptune shows the same behaviour this theory has had to be given up. With Neptune, it seems that the dynamo electric currents responsible for the magnetic field must be close to the surface; the field strength at the planet's surface ranges from 1.2 gauss in the southern hemisphere to a minimum of 0.06 in the northern.

As Voyager moved in, details on the Neptunian surface became striking. The planet is indeed blue, but it is by no means devoid of detail. There is one huge dark spot with an area equal to the entire surface of the Earth; it has become known as the Great Dark Spot (see Figure 3), and has a rotation period appreciably longer than its surroundings. Above it are bright cirrus-type clouds – not of water vapour, but of methane crystals; their average height is some 30 miles above the main cloud deck, and between them and the Great Dark Spot there is what appears to be a clear layer. Among other prominent features are a smaller dark spot, further south on the planet, which has a quicker rotation, and a cloud-like feature which has been nicknamed 'the Scooter' because its rotation period is faster still; every few Neptunian days it catches up in longitude with the Great Dark Spot and 'laps' it. Winds are violent, and may

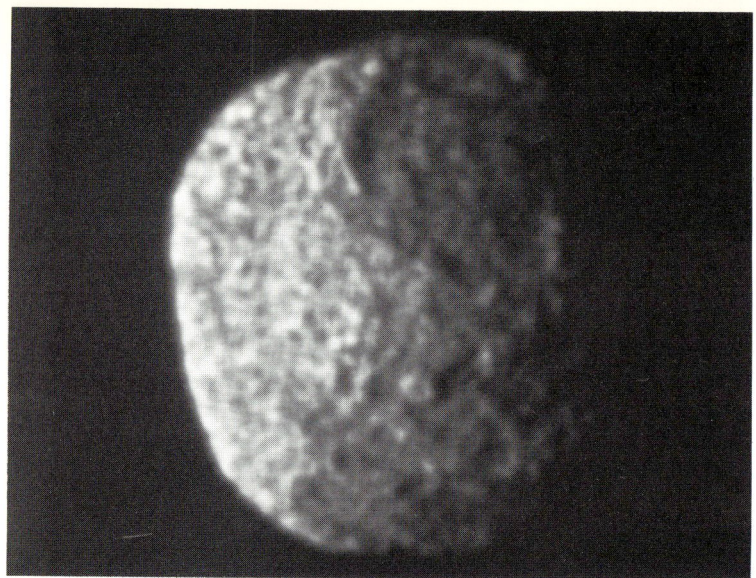

Figure 1. New Satellite (Nl). August 25, 1989 from 146,000 km (91,000 miles). The apparent graininess of the image is caused by the short exposure necessary to avoid smearing. The satellite is dark, and irregular in shape.

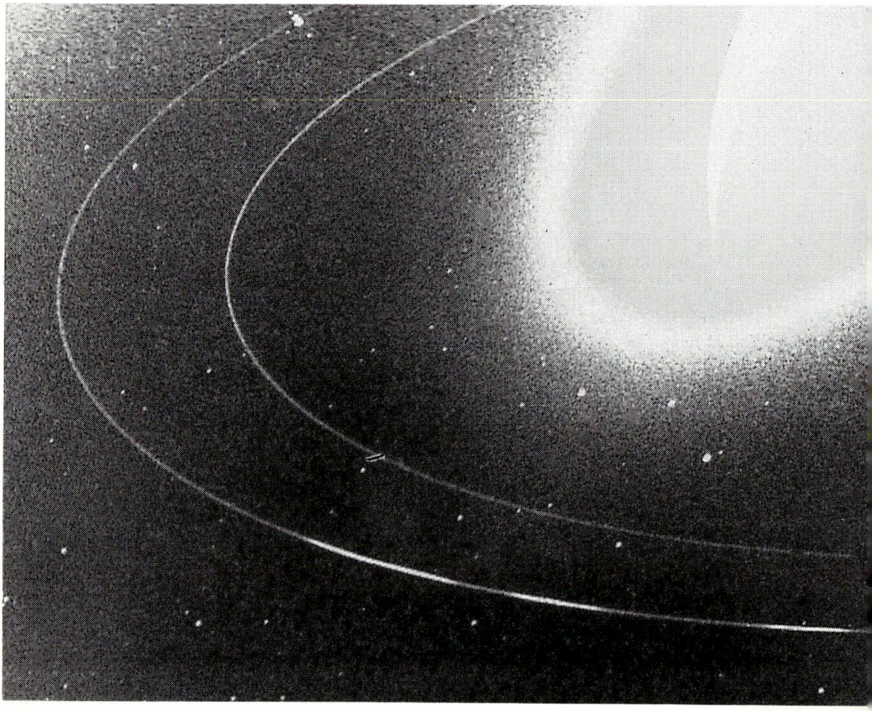

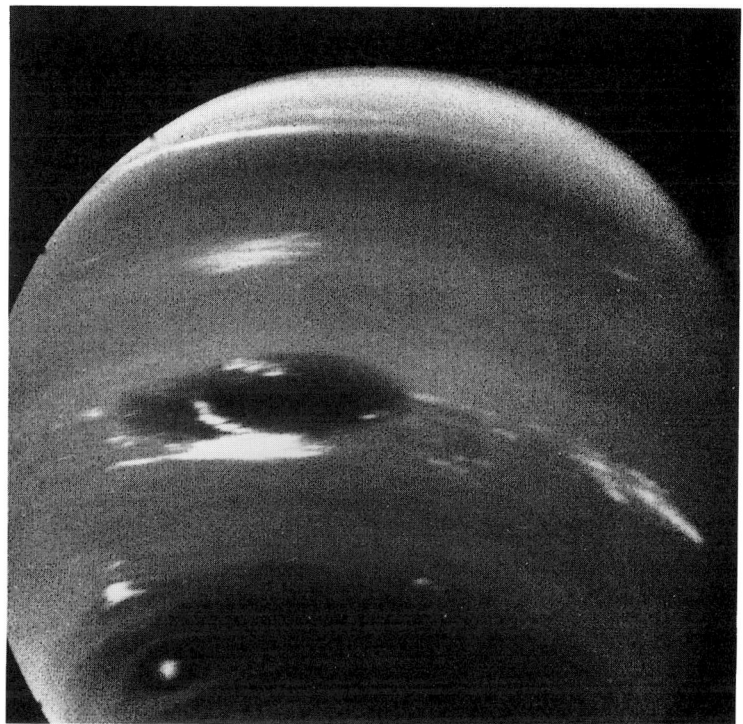

Figure 3. Neptune, as imaged from Voyager 2 on August 21, 1989 from a range of 3,800,000 miles. The Great Dark Spot, with its bright cirrus clouds, is slightly to the left of centre. 'The Scooter' is below and to the left, while the second dark spot, with its bright core, is below the Scooter.

Figure 2 (Opposite). The Rings of Neptune. Wide-angle Voyager 2 image, August 26, 1989. This was the first image to show the rings in detail. The two main rings are 53,000 km (33,000 miles) and 63,000 km (39,000 miles) from Neptune; they are here seen in backscattered light. At this time Voyager was 1,100,000 km (683,000 miles) from Neptune; exposure time was 11 seconds.

reach 700 miles per hour, so that Neptune is anything but placid.

Neptune's hydrogen-rich atmosphere contains methane clouds which show a fairly definite cycle:

(1) Solar ultra-violet radiation destroys the methane in the high atmosphere and converts it to hydrocarbons, such as ethane and acetylene.

(2) The ethane and acetylene descend to the colder, lower stratosphere, where they evaporate and condense.

(3) The hydrocarbon ice particles fall into the warmer troposphere, evaporate, and are converted back to methane.

(4) Buoyant, convective methane clouds rise back to the stratosphere, so returning methane to the upper regions and preventing any net loss.

In the early morning of August 25, Voyager 2 skimmed over Neptune's darkened north pole at a height of only 3000 miles – closer than it had ever been to Jupiter, Saturn, or Uranus; it was then 2,750,876,750 miles from Earth. It had been in space for twelve years, and had covered 4.4 thousand million miles – and yet it reached its target within six minutes of the predicted time.

Crossing the ring-plane was a tense moment. By that time nobody really expected damage from particles or radiation belts, but there was a brief period when the impact rate of minute particles became somewhat alarming, and it was a relief when the rings had been left behind. Triton, the next and last target, was encountered just over five miles after the passage over Neptune's pole.

Triton had been almost completely unknown. Its diameter had been given as anything between 2000 and 6000 miles, and there were suggestions that its atmosphere – believed to be of methane – might be dense enough and cloudy enough to hide the surface, as had been the case with Titan, the largest satellite of Saturn. Oceans of liquid nitrogen or methane had also been proposed. In fact, it was found that the atmosphere is very thin, with a ground pressure of only 0.01 of a millibar, as against about 1000 millibars for Earth and around 6 millibars for Mars, the main constituent is nitrogen, with some methane at lower levels. Moreover, Triton was smaller than had been expected, with a diameter of only 1690 miles – considerably less than that of our Moon though still a little larger than Pluto. Since it was smaller than anticipated, it also had to be more reflective – and colder.

The surface proved to be amazingly varied, with regions which were pinkish and others which looked blue. The very low temperature, −400 degrees Fahrenheit or −236 degrees Centigrade, ruled out oceans. Triton is the coldest place ever encountered by a spacecraft. Instead, much of the surface is coated with a layer of nitrogen ice, with some methane ice mixed in. Surface relief nowhere amounts to much more than three-quarters of a mile, but there are low, irregular formations in some areas, with strange, shallow cracks in others; impact craters are very rare indeed. Of special interest are what have been called 'frozen lakes', and these may give a clue as to Triton's past history (see Figure 4).

All the evidence indicates that Triton is not a true satellite of Neptune. This had long been suspected because, uniquely among large planetary satellites, it moves in a retrograde orbit. What may have happened is that Triton was captured by Neptune in the remote past, and forced into an eccentric path round the planet. Gradually this path became circular, and during the process 'tidal flexing' caused the interior to heat up; a mixture of water and ammonia flowed out on to the surface, filling the caldera-like structures which we see today. Parts of the surface must then have been 'mushy', though now they are solid in the intense cold.

The most surprising idea of all is that Triton may have what may be called 'ice volcanoes'. This suggestion, first made by one of NASA's leading planetary geologists, Dr Laurence Soderblom, was at first regarded as 'crazy' (to use Dr Soderblom's own term) but now appears very probable. Nitrogen ice is abundant on the surface, but not far below – at a depth of, say, 60 to 100 feet – the pressure will be enough to make the nitrogen liquid. If this liquid nitrogen percolates upward, it will reach a region where the pressure is about one-tenth of that of the Earth's air at sea level – and the nitrogen will literally explode in a shower of ice and gas, coming out of the vent in the manner of a geyser at a speed of over 700 feet per second and rising to an altitude of 30 miles or so before falling back. Winds in the thin Tritonian atmosphere will blow it along, producing dark streaks which we can see. The activity could be either continuous or intermittent, but it may well be going in even now, which is strange indeed.

The encounter with Triton marked the end of Voyager's planetary surveys, simply because there are no more known planets. (Pluto could not be reached, as it was in the wrong place, and in any case it is at present closer-in than Neptune; it reached perihelion on

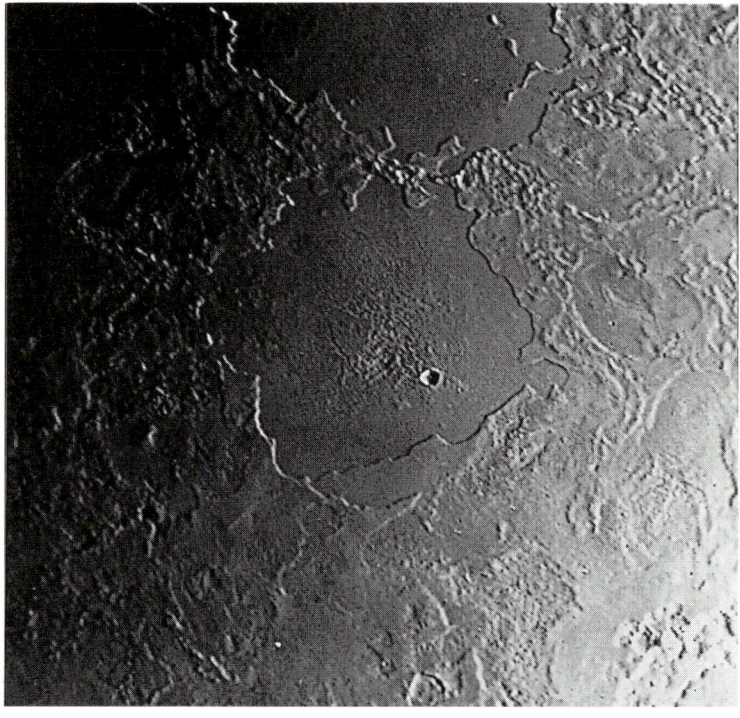

Figure 4. 'Frozen Lake' on Triton. August 25, 1989. The view is about 500 km (300 miles) across. The resolution is about 900 metres (2700 feet).

September 5, 1989.) But Voyager has not reached the end of its useful career. Both it and its twin, Voyager 1, should remain in contact for another twenty-five years or so, by which time they will have reached the boundary of that region of the Galaxy over which the Sun's influence is dominant. Until their power fades, they will go on monitoring such things as the interplanetary magnetic field, the speed and nature of the solar wind, and the numbers of tiny particles in space.

Finally, of course, they will be lost to us. Unless some disaster overtakes it, Voyager 2 will come within 1.7 light-years of the dim red dwarf star Ross 248 in just over 40,000 years from now; in 296,000 years it will be only 4.3 light-years from Sirius – about the same distance as we are from the nearest bright star, Alpha Centauri. We will never know its final fate; perhaps it will be collected by some alien civilization which will puzzle over the recordings which it carries, and will try to decide whence it came. But whether this happens or not, Voyager will never be forgotten by the people of Earth.

All illustrations are reproduced by courtesy of NASA.

Some Interesting Variable Stars

The following stars are of interest for many reasons. The positions are given for epoch 2000. Of course, the periods and ranges of many variables are not constant from one cycle to another.

Star	R.A.		Declination		Range	Type	Period	Spectrum
	h	m	deg.	min.			days	
R Andromedæ	00	24.0	+38	35	5.8–14.9	Mira	409	S
W Andromedæ	02	17.6	+44	18	6.7–14.6	Mira	396	S
U Antliæ	10	35.2	−39	34	5.7– 6.8	Irregular	–	N
θ Apodis	14	05.3	−76	48	6.4– 8.6	Semi-reg.	119	M
R Aquarii	23	43.8	−15	17	5.8–12.4	Symbiotic	387	M+Pec
T Aquarii	20	49.9	−05	09	7.2–14.2	Mira	202	M
R Aquilæ	19	06.4	+08	14	5.5–12.0	Mira	284	M
V Aquilæ	19	04.4	−05	41	6.6– 8.4	Semi-reg.	353	N
η Aquilæ	19	52.5	+01	00	3.5– 4.4	Cepheid	7.2	F–G
U Aræ	17	53.6	−51	41	7.7–14.1	Mira	225	M
R Arietis	02	16.1	+25	03	7.4–13.7	Mira	187	M
U Arietis	03	11.0	+14	48	7.2–15.2	Mira	371	M
ε Aurigæ	05	02.0	+43	49	2.9– 3.8	Eclipsing	9892	F
R Aurigæ	05	17.3	+53	35	6.7–13.9	Mira	457	M
R Boötis	14	37.2	+26	44	6.2–13.1	Mira	223	M
W Boötis	14	43.4	+26	32	4.7– 5.4	Semi-reg.	450	M
X Camelopard	04	45.7	+75	06	7.4–14.2	Mira	144	K–M
R Cancri	08	16.6	+11	44	6.1–11.8	Mira	362	M
X Cancri	08	55.4	+17	14	5.6– 7.5	Semi-reg.	195	N
R Canum Ven.	13	49.0	+39	33	6.5–12.9	Mira	329	M
R Canis Maj.	07	19.5	−16	24	5.7– 6.3	Algol	1.2	F
S Canis Min.	07	32.7	+08	19	6.6–13.2	Mira	333	M
R Carinæ	09	32.2	−62	47	3.9–10.5	Mira	309	M
S Carinæ	10	09.4	−61	33	4.5– 9.9	Mira	150	K–M
ZZ Carinæ	09	45.2	−62	30	3.3– 4.2	Cepheid	35.5	F–K
η Carinæ	10	45.1	−59	41	−0.8– 7.9	Irregular	–	Pec.
γ Cassiopeiæ	00	56.7	+60	43	1.6– 3.3	Irregular	–	B
ρ Cassiopeiæ	23	54.4	+58	30	4.1– 6.2	?	–	F–K
R Cassiopeiæ	23	58.4	+51	24	4.7–13.5	Mira	431	M
W Cassiopeiæ	00	54.9	+58	34	7.8–12.5	Mira	406	N
S Cassiopeiæ	01	19.7	+72	37	7.9–16.1	Mira	612	S
R Centauri	14	16.6	−59	55	5.3–11.8	Mira	546	M
S Centauri	12	24.6	−49	26	6.0– 7.0	Semi-reg.	65	N
T Centauri	13	41.8	−33	36	5.5– 9.0	Semi-reg.	60	K–M
δ Cephei	22	29.2	+58	25	3.5– 4.4	Cepheid	5.4	F–G
μ Cephei	21	43.5	+58	47	3.4– 5.1	Irregular?	–	M
S Cephei	21	35.2	+78	37	7.4–12.9	Mira	487	N
o Ceti	02	19.3	−02	59	1.7–10.1	Mira	332	M
W Ceti	00	02.1	−14	41	7.1–14.8	Mira	361	S
R Chamæleontis	08	21.8	−76	21	7.5–14.2	Mira	335	M
T Columbæ	05	19.3	−33	42	6.6–12.7	Mira	226	M
R Comæ Ber.	12	04.0	+18	49	7.1–14.6	Mira	363	M
R Coronæ Bor.	15	48.6	+28	09	5.7–15	Irregular	–	Fp
W Coronæ Bor.	16	15.4	+37	48	7.8–14.3	Mira	238	M
R Corvi	12	19.6	−19	15	6.7–14.4	Mira	317	M

R Crucis	12	23.6	−61	38	6.4– 7.2	Cepheid	6.7	F
R Cygni	19	36.8	+50	12	6.1–14.2	Mira	426	M
χ Cygni	19	50.6	+32	55	3.3–14.2	Mira	407	S
U Cygni	20	19.6	+47	54	5.9–12.1	Mira	462	N
W Cygni	21	36.0	+45	22	5.0– 7.6	Semi-reg.	126	M
SS Cygni	21	42.7	+43	35	8.4–12.4	Dwarf nova	±50	A–G
R Delphini	20	14.9	+09	05	7.6–13.8	Mira	285	M
U Delphini	20	45.5	+18	05	7.6– 8.9	Semi-reg.?	110?	M
EU Delphini	20	37.9	+18	16	5.8– 6.9	Semi-reg.?	59?	M
β Doradûs	05	33.6	−62	29	3.7– 4.1	Cepheid	9.8	F–G
R Draconis	16	32.7	+66	45	6.7–13.0	Mira	245	M
T Eridani	03	55.2	−24	02	7.4–13.2	Mira	252	M
R Fornacis	02	29.3	−26	06	7.5–13.0	Mira	388	N
η Geminorum	06	14.9	+22	30	3.1– 4.2	Semi-reg.	±233	M
ξ Geminorum	07	04.1	+20	34	3.7– 4.1	Cepheid	10.2	F–G
R Geminorum	07	07.4	+22	42	6.0–14.0	Mira	370	S
U Geminorum	07	55.1	+22	00	8.2–14.9	Dwarf nova	±103	M+WD
S Gruis	22	26.1	−48	26	6.0–15.0	Mira	401	M
α Herculis	17	14.6	+14	23	3.0– 4.0	Semi-reg.	±100	M
S Herculis	17	17.3	+35	06	4.6– 5.3	Beta Lyræ	2.1	B+B
U Herculis	16	25.8	+18	54	6.5–13.4	Mira	406	M
R Hydræ	13	29.7	−23	17	4.0–10.0	Mira	390	M
U Hydræ	10	37.6	−13	23	4.8– 5.8	Semi-reg.	450	N
VW Hydri	04	09.1	−71	18	8.4–14.4	Dwarf nova	100	M
R Leonis	09	47.6	+11	25	4.4–11.3	Mira	312	M
R Leonis Min.	09	45.6	+34	31	6.3–13.2	Mira	372	M
R Leporis	04	59.6	−14	48	5.5–11.7	Mira	432	N
δ Libræ	15	01.1	−08	31	4.9– 5.9	Algol	2.3	B
Y Libræ	15	11.7	−06	01	7.6–14.7	Mira	275	M
R Lyncis	07	01.3	+55	20	7.2–14.5	Mira	379	S
β Lyræ	18	50.1	+33	22	3.3– 4.3	Beta Lyræ	12.9	B+A
R Lyræ	18	55.3	+43	57	3.9– 5.0	Semi-reg.	46	M
RR Lyræ	19	25.5	+42	47	7.1– 8.1	RR Lyræ	0.6	A–F
U Microscopii	20	29.2	−40	25	7.0–14.4	Mira	334	M
U Monocerotis	07	30.8	−09	47	6.1– 8.1	RV Tauri	92	F–K
S Monocerotis	06	41.0	+09	54	4 – 5?	Irregular	–	07
T Normæ	15	44.1	−54	59	6.2–13.6	Mira	243	M
R Octantis	05	26.1	−86	23	6.4–13.2	Mira	406	M
S Octantis	18	08.7	−86	48	7.3–14.0	Mira	259	M
RS Ophiuchi	17	50.2	−06	43	5.3–12.3	Recurrent nova	–	O+M
X Ophiuchi	18	38.3	+08	50	5.9– 9.2	Mira	334	M+K
α Orionis	05	55.2	+07	24	0.1– 0.9	Semi-reg.	± 2110	M
U Orionis	05	55.8	+20	10	4.8–12.6	Mira	372	M
W Orionis	05	05.4	+01	11	5.9– 7.7	Semi-reg.	212	N
ϰ Pavonis	18	56.9	−67	14	3.9– 4.7	W Virginis	9.1	F
S Pavonis	19	55.2	−59	12	6.6–10.4	Semi-reg.	386	M
β Pegasi	23	03.8	+28	05	2.3– 2.8	Semi-reg.	38	M
R Pegasi	23	06.8	+10	33	6.9–13.8	Mira	378	M
β Persei	03	08.2	+40	57	2.2– 3.4	Algol	2.9	B+G
ϱ Persei	03	05.2	+38	50	3 – 4	Semi-reg.	33to55	M
X Persei	03	55.4	+31	03	6 – 7	Irreg. (X-ray)	–	09.5
ζ Phœnicis	01	08.4	−55	15	3.9– 4.4	Algol	1.7	B+B
R Pictoris	04	46.2	−49	15	6.7–10.0	Semi-reg.	164	M
L² Puppis	07	13.5	−44	39	2.6– 6.2	Semi-reg.	140	M
T Pyxidis	09	04.7	−32	23	6.3–14.0	Recurrent nova	–	Q
U Sagittæ	19	18.8	+19	37	6.6– 9.2	Algol	3.4	B–K
WZ Sagittæ	20	07.6	+17	42	7.0–15.5	Recurrent nova	–	Q
RR Sagittarii	19	55.9	−29	11	5.6–14.0	Mira	335	M
RT Scorpii	17	03.5	−36	55	7.0–16.0	Mira	449	M
RY Sagittarii	19	16.5	−33	31	6.0–15	R Coronæ	–	Gp
S Sculptoris	00	15.4	−32	03	5.5–13.6	Mira	365	M
R Scuti	18	47.5	−05	42	4.4– 8.2	RV Tauri	140	G–K
R Serpentis	15	50.7	+15	08	5.1–14.4	Mira	356	M
S Serpentis	15	21.7	+14	19	7.0–14.1	Mira	369	M
λ Tauri	04	00.7	+12	29	3.3– 3.8	Algol	3.9	B+A

T Tauri	04	22.0	+19	32	8.4–13.5	T Tauri	–	G–K
SU Tauri	05	49.1	+19	04	9.1–16.0	R Coronæ	–	Gp
R Trianguli	02	37.0	+34	16	5.4–12.6	Mira	266	M
R Ursæ Major.	10	44.6	+58	47	6.7–13.4	Mira	302	M
T Ursæ Major.	12	36.4	+59	29	6.6–13.4	Mira	256	M
U Ursæ Minor.	14	17.3	+66	48	7.4–12.7	Mira	326	M
X Virginis	12	01.9	+09	04	7.3–11.2	?	–	F
SS Virginis	12	25.3	+00	48	6.0– 9.6	Mira	355	N
R Virginis	12	38.5	+06	59	6.0–12.1	Mira	146	M
S Virginis	13	33.0	−07	12	6.3–13.2	Mira	377	M
R Vulpeculæ	21	04.4	+23	49	7.0–14.3	Mira	136	M
Z Vulpeculæ	19	21.7	+25	34	7.4– 9.2	Algol	2.5	B + A

Mira Stars: maxima and minima, 1990

JOHN ISLES

Below are given predicted dates of maxima and minima for Mira stars on the programme of the BAA VSS, together with (usually) the *mean* range (p = photographic, otherwise visual), period (P), and fraction of the period taken in rising from minimum (m) to maximum (M) for each star. All dates are only approximate.

Star	Range M	Range m	Period d	$(M-m)/P$	Max.	Min.
R And	6.9	14.3	409	0.38	July 25	Feb. 19
W And	7.4	13.7	396	0.42	Dec. 31	July 18
RW And	8.7	14.8	430	0.36	Dec. 15	July 13
R Aql	6.1	11.5	284	0.42	Mar. 28	Sep. 9
V Cam	9.9	15.4	522	0.31	July 21	Feb. 10
X Cam	8.1	12.6	144	0.49	Mar. 27,	Jan. 16
					Aug. 18	June 9
						Oct. 31
SU Cnc*	12.0p	16p	187	0.43	June 29	Apr. 10
						Oct. 14
U CVn*	8.8p	12.5p	346	0.39	Mar. 18	Oct. 15
RT CVn*	12.0p	16.0p	254	0.45	Aug. 10	Apr. 18
						Dec. 28
S Cas	9.7	14.8	612	0.43	Sep. 9	–
T Cas	7.9	11.9	445	0.56	Sep. 20	Jan. 13
o Cet	3.4	9.3	332	0.38	Oct. 1	May 28
R Com	8.5	14.2	363	0.38	Dec. 19	Aug. 3
S CrB	7.3	12.9	360	0.35	Nov. 25	July 22
V CrB	7.5	11.0	358	0.41	June 15	Jan. 19
W CrB	8.5	13.5	238	0.45	Apr. 5,	Aug. 14
					Nov. 29	
R Cyg	7.5	13.9	426	0.35	June 7	Jan. 9
S Cyg	10.3	16.0	323	0.50	Mar. 21	Aug. 29
V Cyg	9.1	12.8	421	0.46	Nov. 27	May 17
Chi Cyg	5.2	13.4	408	0.41	–	Aug. 12
T Dra	9.6	12.3	422	0.44	Oct. 11	Apr. 9
RU Her	8.0	13.7	485	0.43	–	June 23
SS Her	9.2	12.4	107	0.48	Apr. 17,	Feb. 23
					Aug. 2	June 12
					Nov. 17	Sep. 27

Star	Range		Period	$(M-m)/P$	Max.	Min.
	M	m	d			
R Hya	4.5	9.5	389	0.49	Nov. 26	May 19
SU Lac*	11.3p	16p	302	0.40	Oct. 24	June 25
RS Leo*	10.7p	16.0p	208	0.31	Feb. 6, Sep. 2	June 30
W Lyn*	7.5	14.0	295	0.47	Jan. 28, Nov. 19	July 4
X Lyn*	9.5	16	321	0.43	Feb. 2, Dec. 20	Aug. 4 Nov. 16
X Oph	6.8	8.8	329	0.53	June 15	
U Ori	6.3	12.0	368	0.38	Dec. 6	July 19
R Ser	6.9	13.4	356	0.41	Apr. 22	Nov. 18
T UMa	7.7	12.9	257	0.41	Jan. 3, Oct. 18	July 5

*extreme range is given.

Some Interesting Double Stars

We are very grateful to Robert Argyle for this revised list of double stars, which is up to date.

Name	Magnitudes	Separation "	Position angle °	Remarks
Gamma Andromedæ	3.0, 5.0	9.4	064	Yellow, blue. B is again double (0″.5) but needs larger telescope.
Zeta Aquarii	4.4, 4.6	1.8	217	Becoming more difficult.
Gamma Arietis	4.2, 4.4	7.8	000	Very easy.
Theta Aurigæ	2.7, 7.2	3.5	313	Stiff test for 3″0G.
Delta Boötis	3.2, 7.4	105	079	Fixed.
Epsilon Boötis	3.0, 6.3	2.8	335	Yellow, blue. Fine pair.
Kappa Boötis	5.1, 7.2	13.6	237	Easy.
Zeta Cancri	5.6, 6.1	5.6	085	Again double.
Iota Cancri	4.4, 6.5	31	307	Easy. Yellow, blue.
Alpha Canum Ven.	3.2, 5.7	19.6	228	Easy. Yellowish, bluish.
Alpha Capricorni	3.3, 4.2	376	291	Naked-eye pair.
Eta Cassiopeiæ	3.7, 7.4	12.2	310	Easy. Creamy, bluish.
Beta Cephei	3.3, 8.0	14	250	Easy with a 3 in.
Delta Cephei	var, 7.5	41	192	Very easy.
Alpha Centauri	0.0, 1.7	21.7	212	Very easy. Binary, period 80 years.
Xi Cephei	4.7, 6.5	6.3	270	Reasonably easy.
Gamma Ceti	3.7, 6.2	2.9	294	Not too easy.
Alpha Circini	3.4, 8.8	15.7	230	PA slowly decreasing.
Zeta Coronæ Bor.	4.0, 4.9	6.3	305	PA slowly increasing.
Delta Corvi	3.0, 8.5	24	214	Easy with 3 in.
Alpha Crucis	1.6, 2.1	4.7	114	Third star in low-power field.
Gamma Crucis	1.6, 6.7	111	212	Wide optical pair.
Beta Cygni	3.0, 5.3	34.3	055	Glorious. Yellow, blue.
61 Cygni	5.3, 5.9	29	147	Slowly widening. (Add .5)
Gamma Delphini	4.0, 5.0	9.6	268	Easy. Yellow, greenish.
Nu Draconis	4.6, 4.6	62	312	Naked-eye pair.
Alpha Geminorum	2.0, 2.8	2.6	085	Becoming easier.
Delta Geminorum	3.2, 8.2	6.5	120	Not too easy.
Alpha Herculis	var, 6.1	4.6	106	Red, green.
Delta Herculis	3.0, 7.5	8.6	262	Optical pair.
Zeta Herculis	3.0, 6.5	1.5	110	Fine, rapid binary (34y)
Gamma Leonis	2.6, 3.8	4.4	123	Binary; 619 years.
Alpha Lyræ	0.0, 10.5	73	180	Optical. B is faint.
Epsilon Lyræ	4.6, 6.3	2.6	356	Quadruple. Both pairs
	4.9, 5.2	2.2	093	separable with 3 in.
Zeta Lyræ	4.2, 5.5	44	149	Fixed. Easy double.

Name	Magnitudes	Separation "	Position angle °	Remarks
Beta Orionis	0.1, 6.7	9.5	205	Can be split with 3 in.
Iota Orionis	3.2, 7.3	11.8	141	Enmeshed in nebulosity.
Theta Orionis	6.8, 7.9	8.7	032	Trapezium in M. 42.
	6.8, 5.4	13.4	241	
Sigma Orionis	4.0, 10.3	11.1	236	Quadruple. C is rather
	6.8, 8.0	30.1	231	faint in small apertures.
Zeta Orionis	2.0, 4.2	2.4	162	Can be split with 3 in.
Eta Persei	4.0, 8.5	28.5	300	Yellow, bluish.
Beta Phœnicis	4.1, 4.1	1.1	319	Slowly closing.
Beta Piscis Austr.	4.4, 7.9	30.4	172	Optical pair. Fixed.
Alpha Piscium	4.3, 5.3	1.9	283	Binary; 720 years.
Kappa Puppis	4.5, 4.6	9.8	318	Again double.
Alpha Scorpii	0.9, 6.8	3.0	275	Red, green.
Nu Scorpii	4.2, 6.5	42	336	Both again double.
Theta Serpentis	4.1, 4.1	22.3	103	Very easy.
Alpha Tauri	0.8, 11.2	131	032	Wide, but B very faint in small telescopes.
Beta Tucanæ	4.5, 4.5	27.1	170	Both again double.
Zeta Ursæ Majoris	2.1, 4.2	14.4	151	Very easy. Naked-eye pair with Alcor.
Alpha Ursæ Minoris	2.0, 9.0	18.3	217	Can be seen with 3 in.
Gamma Virginis	3.6, 3.7	3.5	292	Binary; 171 years. Closing.
Theta Virginis	4.0, 9.0	7.1	343	Not too easy.
Gamma Volantis	3.9, 5.8	13.8	299	Very slow binary.

Some Interesting Nebulæ and Clusters

Object	R.A.		Dec.		Remarks
	h	*m*			
M.31 Andromedæ	00	40.7	+41	05	Great Galaxy, visible to naked eye.
H.VIII 78 Cassiopeiæ	00	41.3	+61	36	Fine cluster, between Gamma and Kappa Cassiopeiæ.
M.33 Trianguli	01	31.8	+30	28	Spiral. Difficult with small apertures.
H.VI 33–4 Persei	02	18.3	+56	59	Double cluster; Sword-handle.
△142 Doradûs	05	39.1	−69	09	Looped nebula round 30 Doradûs. Naked-eye. In Large Cloud of Magellan.
M.1 Tauri	05	32.3	+22	00	Crab Nebula, near Zeta Tauri.
M.42 Orionis	05	33.4	−05	24	Great Nebula. Contains the famous Trapezium, Theta Orionis.
M.35 Geminorum	06	06.5	+24	21	Open cluster near Eta Geminorum.
H.VII 2 Monocerotis	06	30.7	+04	53	Open cluster, just visible to naked eye.
M.41 Canis Majoris	06	45.5	−20	42	Open cluster, just visible to naked eye.
M.47 Puppis	07	34.3	−14	22	Mag. 5,2. Loose cluster.
H.IV 64 Puppis	07	39.6	−18	05	Bright planetary in rich neighbourhood.
M.46 Puppis	07	39.5	−14	42	Open cluster.
M.44 Cancri	08	38	+20	07	Præsepe. Open cluster near Delta Cancri. Visible to naked eye.
M.97 Ursæ Majoris	11	12.6	+55	13	Owl Nebula, diameter 3'. Planetary.
Kappa Crucis	12	50.7	−60	05	'Jewel Box'; open cluster, with stars of contrasting colours.
M.3 Can. Ven.	13	40.6	+28	34	Bright globular.
Omega Centauri	13	23.7	−47	03	Finest of all globulars. Easy with naked eye.
M.80 Scorpii	16	14.9	−22	53	Globular, between Antares and Beta Scorpionis.
M.4 Scorpii	16	21.5	−26	26	Open cluster close to Antares.
M.13 Herculis	16	40	+36	31	Globular. Just visible to naked eye.
M.92 Herculis	16	16.1	+43	11	Globular. Between Iota and Eta Herculis.
M.6 Scorpii	17	36.8	−32	11	Open cluster; naked eye.
M.7 Scorpii	17	50.6	−34	48	Very bright open cluster; naked eye.
M.23 Sagittarii	17	54.8	−19	01	Open cluster nearly 50' in diameter.
H.IV 37 Draconis	17	58.6	+66	38	Bright Planetary.
M.8 Sagittarii	18	01.4	−24	23	Lagoon Nebula. Gaseous. Just visible with naked eye.
NGC 6572 Ophiuchi	18	10.9	+06	50	Bright planetary, between Beta Ophiuchi and Zeta Aquilæ.
M.17 Sagittarii	18	18.8	−16	12	Omega Nebula. Gaseous. Large and bright.
M.11 Scuti	18	49.0	−06	19	Wild Duck. Bright open cluster.
M.57 Lyræ	18	52.6	+32	59	Ring Nebula. Brightest of planetaries.
M.27 Vulpeculæ	19	58.1	+22	37	Dumb-bell Nebula, near Gamma Sagittæ.
H.IV 1 Aquarii	21	02.1	−11	31	Bright planetary near Nu Aquarii.
M.15 Pegasi	21	28.3	+12	01	Bright globular, near Epsilon Pegasi.
M.39 Cygni	21	31.0	+48	17	Open cluster between Deneb and Alpha Lacertæ. Well seen with low powers.

Our Contributors

Dr David Allen, of the Siding Spring Observatory in New South Wales, needs no introduction; he is our most regular contributor, and the *Yearbook* would be incomplete without him! He continues his researches, largely in the realm of infrared astronomy, as well as writing and broadcasting.

Jonathan Davies, a Ph.D. student in the Department of Physics, University of Wales College of Cardiff, is currently writing his thesis on the subject of low surface brightness galaxies.

Michael De Faubert Maunder, a Member of the Council of the British Astronomical Association, is perhaps best known as a skilled astronomical photographer – and has been on several eclipse expeditions.

Professor Michael Disney, who took his Ph.D. from the University of London, is now Professor of Astronomy at the University of Wales College of Cardiff. He was co-discoverer of the first optical pulsar, and is a member of the Hubble Space Telescope Faint Object Camera team.

Dr Patrick Moore, Editor of this *Yearbook*, was President of the British Astronomical Association for the 1982–84 sessions. He is Editor-in-Chief of the British monthly periodical *Astronomy Now*.

Dr Paul Murdin, one of the world's leading astrophysicists, is at the Royal Greenwich Observatory, Cambridge. He has been in charge of the British telescopes at La Palma. He is the author of popular books as well as many technical papers, and is a frequent broadcaster on sound and television. He is also a specialist in studies of supernovæ.

Dr Steven Phillipps, who took his Ph.D. from the University of Durham, is now a lecturer in the Department of Physics, University of Wales College of Cardiff.

Gordon Taylor is, of course, the author of the lists of monthly phenomena in the *Yearbook*; this year he has also contributed a main article. He was for many years a member of the staff of the Royal Greenwich Observatory, and has made many important contributions to astronomical sciences.

The William Herschel Society maintains the museum now established at 19 New King Street, Bath – the only surviving Herschel house. It also undertakes activities of various kinds. New members would be welcome; those interested are asked to contact Dr L. Hilliard at 2 Lambridge, London Road, Bath.